# DISCOVERING **DINOSAURS** IN THE OLD WEST

# DISCOVERING **DINOSAURS** IN THE OLD WEST

## The Field Journals of Arthur Lakes

Edited by Michael F. Kohl and John S. McIntosh

with a foreword by John Ostrom

Smithsonian Institution Press
Washington and London

Editor: Jack Kirshbaum
Designer: Linda McKnight

Library of Congress Cataloging-in-Publication Data

Discovering dinosaurs in the Old West : the field journals of
Arthur Lakes / edited by Michael F. Kohl and John S. McIntosh.
p. cm.
Includes bibliographical references and index.
ISBN 1-56098-700-6 (cloth : alk. paper)
1. Dinosaurs—Colorado—Morrison Region.
2. Dinosaurs—Wyoming—Como Bluff. 3. Lakes, Arthur,
1844–1917—Diaries. I. Kohl, Michael F. II. McIntosh,
John Stanton, 1923–
QE862.D5D528 1997
567.9′09787′86—dc21 97-20616

British Library Cataloging-in-Publication data available

Manufactured in the United States of America
02 01 00 99 98 97 5 4 3 2 1

∞ The paper used in this publication meets the minimum requirements of the American National Standard for Permanence of Paper for Printed Library Materials Z39.48-1984.

Cover: Detail from an 1879 watercolor by Arthur Lakes of W. H. Reed (shown) and E. Kennedy removing bones of *"Brontosaurus" excelsus* from Quarry no. 10. Courtesy of the Peabody Museum of Natural History, Yale University.

Frontispiece: Arthur Lakes. Courtesy of the Arthur Lakes Library, Colorado School of Mines.

To our parents, who fostered a love of science
and a wonder at a world where dinosaurs once lived.

# Contents

# Foreword

Exploration of the western frontier must have been one of the most exciting adventures in American history. Carried out over the centuries by successive generations of early American pioneers from all walks of life, these were the historians of our legacy. Fortunately for us, much of that adventure was recorded and passed on for those that followed. Often these "memos from the past" were registered by well-educated and sometimes very talented explorers who left us vivid descriptions of their experiences, places, and events. The paintings left by Arthur Lakes were executed during his explorations and teaching excursions in the wilderness of the Colorado and Wyoming territories and are well known to many twentieth-century geologists and paleontologists. But his written journals are less well known, and some have come to light only recently, filling in the unpainted details of Lakes's nineteenth-century frontier world.

Professor Arthur Lakes taught in a local mining school in Golden, Colorado (which became the Colorado School of Mines), and like most earth scientists of today he spent much of his time exploring the wilderness, the remote terrain in his expanding frontier. His innate ability to "read" the geology around him and his ingrained compulsion to sketch and paint the scenes before him made him the ideal field observer, one of remarkable value to his mining school students and to his correspondents. As a naturalist, he was a highly skilled recorder before the camera or audio recorders came into wide use as essential field equipment for the field explorers and scientists.

In April 1877 Lakes wrote to Professor Othniel Charles Marsh of Yale

College, one of the New World's recognized academic centers of earth science education. Marsh was prominent in the field of paleontology and one of few experts available for consultation at that time. In the course of his letter to Marsh, Lakes said: "A few days ago whilst taking a geological section and measurements - and examining the rocks of Bear Creek near the little town of Morrison about fifteen miles west of Denver, I discovered in company with a friend, a Mr. Beckwith of Connecticut, some enormous bones apparently a vertebra and a humerus bone of some gigantic saurian in the Upper Jurassic or Lower Cretaceous at the base of Hayden's Cretaceous No. 1 Dakotah group" (*Othniel Charles Marsh Papers,* reel 10, frame 355ff).

This letter, answered belatedly by Marsh, triggered a very fruitful collaboration that endured for a number of years. It led to Lakes's collection of the earliest Morrison formation dinosaurs which he recorded in reports and sketches. In July 1877 Marsh learned about the discovery of other similar giant fossil bones near Laramie, Wyoming, at Como Bluff. By the spring of 1879, he redirected his collectors to Wyoming, where Lakes met the indefatigable collector and loyal worker, William Harlow Reed.

This quite ordinary-looking locale in southern Wyoming had an extraordinary influence on the development, both philosophical and architectural, of many of the great museums of the New World, and probably on the principal museums of the world as well. From Como Bluff came the first major discoveries of dinosaur remains anywhere in the world. From this place were collected many of the fine skeletons now displayed in the Peabody Museum at Yale University in New Haven, Connecticut, in the National Museum of Natural History at the Smithsonian Institution in Washington, D.C., and in the American Museum of Natural History in New York City.

By the 1900s it was fashionable to design museum buildings around the prospect or specific plan that such structures would in due time shelter an assortment of giant saurians of past eras. And it was not long before there was a brisk trade in dinosaur skeletons to many of the smaller museums being founded all over North America and to many of the well-established and famous museums of Europe. Dinosaur skeletons became status symbols and soon every museum had to have its own; today they are still very much in fashion. All of this can be attributed directly to Como Bluff and the great variety, the unbelievable numbers, and the exceptional preservation of dinosaurian skeletons uncovered there. Never before had such spectacular fossil remains been found or in such abundance.

Fragmentary dinosaur bones had been found at other localities in England and western Europe, and in eastern America in the early 1800s, but it was at Morrison, Colorado, and soon after at Como Bluff, Wyoming, that the first "dinosaur rush" began. And it was here at Morrison that Arthur Lakes began recording his discoveries in his journal for future generations.

Professor John Ostrom
Peabody Museum of Natural History
Division of Vertebrate Paleontology
Yale University

# Preface

## Project Inception

During the summer of 1994, I consulted a number of primary sources documenting the history of paleontology for a presentation at the Society of American Archivists meeting. Among the documents at the Smithsonian Institution Archives were two journals of Arthur Lakes, one describing his fieldwork at Morrison, Colorado, and the other his work at Como Bluff, Wyoming (listed as record unit 7201 in the *Guide to the Smithsonian Archives* [Washingon, D.C.: Smithsonian Institution Press, 1978]). Assuming that the journals had been published, I contacted Dr. Mike Brett-Surman, of the Smithsonian's Paleobiology Department, with whom I had worked as a volunteer on a dig in Wyoming, and asked for the citation. Brett-Surman suggested that contacting Professor Jack McIntosh of Wesleyan University, who had coauthored a book about one of the dig sites. I was astonished when McIntosh called and stated that the Lakes journals had never been published and were assumed to have been lost because they were not with the records of O. C. Marsh at Yale. His evaluation that the journals were a major resource documenting the discovery of dinosaurs in the West, and his enthusiastic encouragement that they be edited, formed the basis for this collaborative project.

## Provenance

From scattered internal evidence, it is likely that parts of the journals were actually written sometime after the events described. In the Morrison journal, Lakes mentions that some of the fossils were identified as turtles by Marsh which could have occurred only after the material had been shipped to New Haven. Both journals contain material at their end that are of a later date. The Morrison journal concludes with events that occurred in 1878. In a portion of the Como journal, the term *Baptanodon* is used as well as *Sauranodon* for the ichthyosaurs discovered at Como Bluff. Marsh only renamed them in late 1880, after he realized that the name *Sauranodon* was already used for the name of another creature. The similarity between portions of the journal and Lakes's letters and journal reports found in the Marsh Papers, however, does demonstrate that the journals were generally written soon after the events described.

Sometime during the 1880s, the journals came to New Haven. This may have occurred at the time when Lakes contacted Marsh regarding a scientific article about mountain rats or they may have been transferred when Lakes sold watercolors of some of his sketches to Marsh in 1889. In any event, they were part of the seven railroad boxcar loads of material transferred to the United States National Museum (now the National Museum of Natural History, Smithsonian Institution) in 1899. This was the settlement of claims related to specimens collected by O.C. Marsh while working for the United States Geological Survey. The Morrison journal has pasted in its inner front cover a note from Frederick Augustus Lucas, a curator at the Smithsonian from 1883 to 1903, indicating that he had read the journal and was identifying the correct date (1877 instead of 1878), when the discoveries at Morrison occurred. Lucas had been in charge of the packing and transfer of material from New Haven, which took months to accomplish.

The volumes were accessioned by the Smithsonian Library on January 19, 1900, accession numbers 167085–167088. Three of the volumes were catalogued for the manuscript case: *Journal of Travel and Exploration Discovery of Dinosaurs at Morrison, Journal of Explorations for Saurians and Fossil Remains in Wyoming,* and *Notes and Lectures on the Reptilian Age and Discoveries of Dinosaurs*. In the early 1970s these three volumes were transferred to the Smithsonian Institution Archives and catalogued as record unit 7201.

One volume, *Explorations for and Discoveries of Saurian Remains in Colorado and Wyoming under Direction of Professor Marsh of Yale during 1878—'79—'80,* a 108-page album of pen and ink sketches was catalogued for the Stratigraphic Paleontology Department of the Smithsonian sometime during the first half of the twentieth century.[1] Lakes mentions this sketchbook in his journal and indicates that it contained a variety of subject matter, no doubt including stratigraphic sections. This may explain why it was separated from the other journals, as would the possibility that it is oversize. Lakes sent Marsh a partial listing of its contents in a letter dated May 19, 1888, and it is possible that he produced some of the watercolors that the Peabody Museum now has from sketches in this volume, which was turned over to Marsh at the time of payment of the sketches in February 1889. The volume was not accessioned into the Smithsonian Archives and currently is missing.

## On Method and Style

The goal of this volume is to appeal to the general reader of science and history as well as to professional paleontologists. Its object is to entertain by accurately conveying the excitement of the discovery of dinosaurs as recorded by Arthur Lakes. His journals describe the adventure of the Old West, where train robbers plied their trade and Native Americans fought battles against encroachment in a climate ranging from spring's williwaw to winter's blizzard. Lakes's descriptions of paleontological discovery rank with those written by Roy Chapman Andrews and Roland T. Bird.

Editorial decisions are guided by the needs of this audience. We attempt to reproduce the journals faithfully with a minimum of editorial intrusion but at the same time to clarify when necessary journals created for the author's personal use. Spelling has been Americanized, and abbreviations are spelled out, as are numbers in the text under one hundred. Lakes abbreviated words ending in "ing" or "ly" and these words have also been spelled out. He also used the double ss symbol, ß, which is not employed here. Punctuation is added to entries where few breaks in the narrative are noted other than the capitalization of the next word. Lakes's additions, words in superscript and subscript, and words crossed out have not been noted unless they have some particular significance. Based upon the use of different inks and pencil, it is clear that Lakes made some editorial corrections to the Morrison journal, mainly underlining titles, adding scientific names for creatures, and occasionally adding a word or two. In a number of instances, such as "vertebra" and "stratum," the noun has been changed to agree with the verb. We use brackets to indicate comments added by us, usually a short identification or definition, as well as when a word has been added. Abbreviations have been used for the Yale

Peabody Museum (YPM) and the American Museum of Natural History (AMNH) when identifying particular specimens. Scholars are encouraged to consult the journals themselves regarding these aspects of Lakes's composition.

The journals themselves present a problem with respect to the identification of what constitutes a unit of entry, such as would be the case with a letter or an essay in a different editorial project. In the case of the Colorado journal, there are no page numbers and almost every page begins with a title although specific dates for entries are infrequent and incomplete. The Wyoming journal presents the contrast of having prenumbered pages and many more dated entries but few page titles. It does have frequent headings in the margin. Only some of these titles and marginal headings have been retained. In some instances a heading has been moved to where it more appropriately fits in the narrative.

The Wyoming journal has the additional feature of containing copied portions of letters that Lakes wrote to his family in England during May and June 1879. There is no mention of family matters and one supposes that he copied parts of the letters to further document what he did not include in his journal. This copied portion repeats a fair amount of material found in his journal but includes extensive additional information about the personalities of Lakes's fellow companions at Como Bluff. Material from the letters to England has been added to the text where it supplements the journal entries and is identified as such. There is no explanation as to why the copying of the letters stopped, although Lakes spent July 1879 through March 1880 at Como Bluff.

This editorial project has been confined specifically to the two journals Arthur Lakes wrote concerning his work collecting fossils for O.C. Marsh in Colorado and Wyoming. A third journal at the Smithsonian Archives, containing research and lecture notes about geology and paleontology, has not been edited because its contents, although related by subject, are not in a narrative, chronological format and pertain to general scientific information current in the 1880s. Lakes's extensive correspondence with Marsh is available for use by researchers at the Sterling Library of Yale University. The Marsh Papers have been microfilmed and can be purchased. Lakes correspondence is also scattered in collections at the Smithsonian as well as two survey field notebooks at the National Archives from when he worked for the USGS assisting in the survey of the Leadville District and Mosquito Mountains.

## Acknowledgments

The editors would like to thank Mike Brett-Surman and the members of the Paleobiology Department of the Natural History Museum of the Smithsonian for their help and hospitality at all stages from this project's inception. The hospitality and help of Mary Ann Turner and Barbara Narendra of the Peabody Musuem's is also appreciated, particularly with respect to Lakes's paintings and correspondence. The enthusiasm and support of John Ostrom of Yale University helped ensure that this publication saw the light of day.

We would like to thank all the archivists and librarians without whose help this project would have been impossible: those at the Smithsonian Institution's Archives and Libraries who helped to track down the provenance of these journals and continue to search for the missing sketchbook volume, including Alan Bain, William Cox, William Deiss, John Fleckner, James Glenn, Susan Glenn, Edie Hedlin, Paulette Hughes, Jose Jamison, Lesley Overstreet, Tammy Peters, Robert Skarr, and Paul Theerman. The reference librarians at the Colorado Historical Society, the Colorado School of Mines, the Denver Public Library, the Library of Congress, the State Historical Society of Wisconsin, the University of Kansas, the University of Wisconsin—Madison, and the University of Wyoming; the archivists at the American Heritage Center of the University of Wyoming, the Spencer Library at the University of Kansas, the National Archives Textual Reference Division, the Naval Historical Center Operations Archives and Yale University's Sterling Library Special Collections. Many members of the Clemson University Libraries supported this project in a great number of ways and ensured its completion in a timely manner. The advice and help of reference and interlibrary librarians and staff to the project deserve a special thanks.

David Norman, Director of the University of Cambridge Sedgwick Museum located information about Lakes's stay at Oxford that was most helpful as was the response of Geoffrey Marshall Provost of The Queens College, Oxford. Ginny Mast of the Colorado School of Mines Geology Museum graciously provided her research notes and advice concerning Lakes's life in Colorado. Pat Monoco and Donna Engard of the Garden Park Paleontology Society in Cañon City, Colorado, enthusiastically gave access to research material gathered from around the country. Carol Edwards of the USGS Field Records Library was kind enough to post a search request about the missing sketchbook on a number of e-mail lists.

The help of Beth Southwell, independent researcher and assistant paleontologist in Wyoming for Dinamation, is greatly appreciated in checking citations in inexcessable sources. Doris Kneuer, volunteer curator, Florissant Fossil Beds National Monument, provided specific information about the history of that site, and Herbert Meyer and Lee Snapp provided slides for illustrations. Dan Grenard's time was appreciated in Cañon City. Brent Breithaupt's warm hospitality in Laramie and tour of Como Bluff was greatly appreciated during a snowy visit in September.

The American Heritage Center of the University of Wyoming's travel grant helped Michael Kohl to use its wonderful collections. He would like to express his appreciation to the individuals who provided hospitality during his research trips: Diane and Peter Boyer, Terry Burk and Glenna Cloud, Patti Hastreiter and Mike Neal, Sally and John Steadman, and Rose and Ron Zaeske.

He would also like to greatly thank those who granted his sabbatical request: Chair of the Libraries, Deborah Babel; Dean of Libraries, Joseph Boykin, Jr; Provost of the University, Charles Jennett; and President of the University, Phillip Prince. He is particularly grateful for the willingness of the staff of Clemson University Libraries' Special Collections to shoulder the burden of being short-staffed during his sabbatical: James Cross, Linda Ferry, Susan Hiott, Adraine Jackson, Roger Leimhuis, Dennis Taylor, and Laurie Varenhorst.

Finally we would like to thank the editors at the Smithsonian Institution University Press, Peter Cannell and Jack Kirshbaum, for their patient work in helping to craft this publication from a holographic manuscript through to a published volume, as well as Linda McKnight for her excellent design.

## Note on Queens Canyon

In Glen Eyrie ~~Canon~~ Queens Caño is cut to a depth of 700 feet through massive red granite and rocks of the lower Silurian. Upon the massive red coarse granite which is sometimes only red granite felspar & quartz without mica or hornblende. About 100 feet above the stream the Silurian rocks lie unconformably. The layers lie evenly one upon the other. The lowest weathering to a light grey with rugged outline above this is a dark reddish brown massive sand stone. The layer above is deeper red but less massive. The top is capped with grey shales. There is 3 miles of Rock thickness from Colorado City to Manitou From the Niobrara limestone to the Silurian and allowing for dips about 12 to 15000 feet thickness of strata

## Notes on Rocks on the Divide

The divide was evidently once the top of high plain we rode along on what we call the plain but it is really on the deep valleys which have been washed down to a hundred feet below the original plain. The mesas are illustrations that it is water that has made mountains by washing out valleys. The lava capping these must have poured out over them from ~~som~~ some source to the north probably through cracks in the earth.

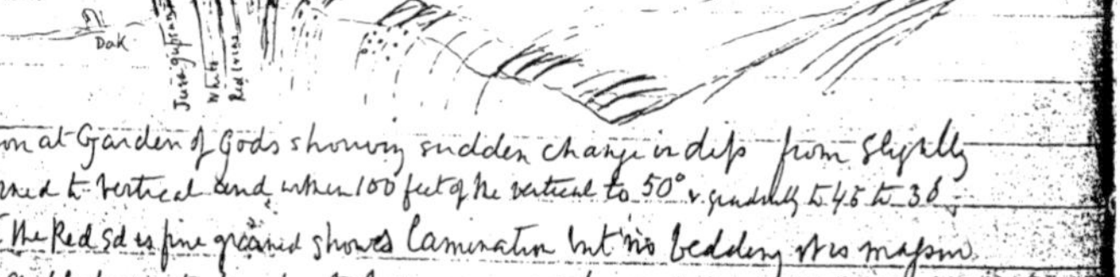

Gate

feet 4100

Oak

Just gypsum

White

Red rocks

Section at Garden of Gods showing sudden change in dip from slightly overturned to vertical and within 100 feet of the vertical to 50° & gradually to 45 to 30. At Garden gate the Red Sd is fine grained shows lamination but no bedding & is massive. There is rather a sudden transition from fine to very coarse conglomerate. and it occurs also where the change of dip is sudden

# Introduction

## United States Paleontology during the 1870s

By the middle of the nineteenth century, paleontology had come to have a key place among the emerging natural sciences because of its relationship with geology and the economic exploitation of mineral wealth as well as its role in providing evidence related to theories about the natural history of the earth and the creatures that inhabited it from the distant past. The earnest and passionate debate regarding how and when life arose on earth was as intellectually stimulating to these scientists and as captivating of the public's attention as nuclear physics and space exploration have been to people in the late twentieth century.

During the 1870s vertebrate paleontology in the United States came to be dominated by two immensely gifted and intensively competitive scientists: Othniel Charles Marsh (1831–99) of Yale University and Edward Drinker Cope (1840–97), who in the 1870s worked for the Hayden Survey. They built upon earlier research done by scientists such as Joseph Leidy (1823–91), who taught at the University of Pennsylvania and then Swarthmore College. Before the 1870s, Leidy had established himself as the best-known American paleontologist. He had described a number of important fossils including, in 1858, the duckbilled dinosaur that he called a *Hadrosaurus*. However by the middle of the 1870s, Leidy had retreated from vertebrate paleontology because of his lack of financial resources and inability to compete with Marsh and Cope, as well as a genuine distaste for their competitive, sometimes unethical, collecting practices. Marsh and Cope may be compared with the captains of industry of this Gilded Age in that they pursued their objectives with persistence unhindered by regulations or much of a sense of propriety. In a number of instances, collectors were hired away by one of the rivals, specimens hidden or smashed, and science ill-served by a rush to name new species based upon fragmentary discoveries.

Marsh and Cope had markedly different personalities, but their backgrounds were similar. Most strikingly, neither had participated in the seminal event of nineteenth-century American history: the Civil War. Both spent their time during those years studying in Europe. Each was independently wealthy and willing to use that wealth to further their scientific research and reputations. Each worked for the federal government's sur-

Othniel Charles Marsh. Courtesy of the Smithsonian Archives.

veys—the United States Geological Survey (USGS) and its predecessors—which helped them to collect specimens and provided protection in unsettled territory. These surveys produced volumes of great scientific importance authored by Cope and Marsh as well as many other scientists such as plant paleontologist Leo Lesquereux (1806–89) and paleoentomologist Samuel H. Scudder (1837–1911).

Although the surveys provided some government support, Marsh and Cope relied upon their own network of independent collectors to scout out the best locations and prospect for them. The sites in Colorado at Morrison and Cañon City were discovered and excavated by amateur collectors who hired themselves to work for either Marsh or Cope. In similar fashion, the discoveries at Como Bluff by William Harlow Reed and William Edward Carlin were the result of amateurs who realized that they had stumbled across something of value to paleontologists. Carlin worked first for Marsh but then worked for Cope and quit the business by the early 1880s. Marsh was extremely fortunate that Reed turned out to be a man of numerous capabilities with a dogged sense of loyalty. He worked for Marsh in one capacity or another until the latter's death in 1899. The uneasy relationship between the amateur collector and the scientist continues to this day, as does occasionally the acrimony among collectors.

Arthur Lakes knew of Marsh and Cope from their publications and it is not surprising that he contacted them when he realized the nature of his discoveries of huge bones at Morrison, Colorado, in the spring of 1877. What followed was one of the most interesting chapters in the history of American paleontology. Lakes documented these events in his journals.

## Arthur Lakes

Arthur Lakes (1844–1917) made some of the most extraordinary discoveries of dinosaur fossils in the American West. A British-born educator who taught for more than two decades at Colorado's School of Mines, he combined a wide-ranging knowledge of natural science with the ability to write and draw. His discoveries and adventures in discovering dinosaurs was but a brief chapter in a long life of teaching and writing that earned him the title of "Father of Colorado Geology."

Born on December 21, 1844, in Martock Somerset, England, to Rev. John and Catherine Lakes, he was educated at Queen Elizabeth College Guernsey. He entered Queens College of Oxford University during the winter term of 1863 and was in residence, with a break in attendance, until the end of the spring term of 1865. He withdrew that year before the fall term. Lakes subsequently emigrated to the United States, possibly via Canada. In 1869 he began teaching writing and drawing at Jarvis Hall, a newly founded Episcopal boys prep school and college in Golden, Colorado Territory. In 1870 a school of mines was added to the college, as was Mathews Hall, an Episcopalian seminary.[1]

Lakes balanced a multitude of talents and interests during the next decade. He studied the geology of Colorado and began collecting fossils for the plant paleontologist, Leo Lesquereux, whose research was published in F. V. Hayden's annual reports of the USGS of the territories during the 1870s. In 1874 Lakes was ordained a deacon in the Episcopal Church and began serving as itinerant minister to a number of mining communities such as Idaho Springs.[2] He continued teaching writing and drawing at Jarvis Hall but sought means to supplement his income during the summer recesses.

On March 26, 1877, Lakes and Henry C. Beckwith, a retired naval officer, were surveying a ridge of tilted strata, called a hogback, near Morrison when they discovered bones of such size that Lakes at once realized their importance. On April 2 Lakes wrote to Marsh at Yale College, describing this discovery and asking whether Marsh would be interested in acquiring it for his collections. Receiving no answer later that month, Lakes sent specimens to both Marsh and Cope, inquiring whether they might help identify them and possibly purchase them. Among the fossils that Lakes helped discover were those later identified as *Apatosaurus ajax* and *Stegosaurus armatus*. Although Cope offered to hire him, Lakes had already agreed to work for Marsh and so he asked Cope to send his specimens on to Marsh. Marsh agreed to pay Lakes $100 per month with the possibility of an increase to $125 for a period that extended until December 1877, when Marsh ordered a cessation of operations.

In early April 1878, Lakes faced a major career crisis when Jarvis and Mathews Halls burned. Lakes was thrown out of work and had to rely upon fossil collecting. That summer Marsh rehired him to work at Morrison, after Lakes suggested that he had other offers that he might be required to accept, and Marsh invited him to visit New Haven. After discussing the matter with his bishop, Lakes agreed to the journey and spent the winter of 1878–79 at Yale examining specimens and working with Marsh and his assistants. He returned to Colorado in late March and began further work at the Morrison quarries. However, in late April, Marsh requested that Lakes close up operations at Morrison and travel to Como Bluff, Wyoming, where another, more extensive series of Jurassic quarries had been in operation since 1877.

In May he arrived in Como Bluff, where he worked with William Harlow Reed. The juxtaposition of the Oxford-educated, ordained Englishman with the largely self-educated, American frontiersman proved to be explosive, culminating in both Reed and Lakes offering their resignations toward the end of August 1879. It is clear that Reed resented the amount of time Lakes spent preparing sketches and watercolors. That summer Reed and others in Marsh's crew were engaged in the strenuous work of excavating the huge skeleton of the *Brontosaurus excelsus* (now known as *Apatosaurus excelsus*) that today graces the main hall of the Yale Peabody Museum. It should be noted that Marsh requested Lakes to prepare these drawings, which have permitted later scientists to find the quarries. The illustrations are a wonderful source for understanding pioneer paleontology in the Old West and are perhaps Lakes's most lasting contribution to his work at Como. A number of these watercolors are reproduced in this book.

Much to Marsh's relief, both Reed and Lakes continued working—but separately—during the following fall and winter. One may assume that the relationship between the two men improved. Lakes demonstrated a level of pluck by continuing to dig in the stegosaurus quarry during the frightful conditions of a Wyoming winter. Perhaps the fact that Lakes won second place in the marksmanship contest after Reed on New Year's Day helped to reestablish a degree of mutual respect between the men.

Nevertheless, in March 1880 Lakes returned to Golden, Colorado, to teach exclusively at the Colorado School of Mines. Jarvis Hall had been relocated to Denver after the fire. Lakes taught geology and directed his research interests to hard rock—mining geology. Perhaps the rigors of quarrying during a Wyoming winter encouraged him to change his interests as perhaps did Marsh's delays in paying wages due and reluctance to have his field staff share in the publication of their discoveries. When in 1877 one of Lakes's Morrison sketches was published, he had to beg Marsh's pardon. Two years later Lakes did write an article for the *Kansas City Review* about the discovery of dinosaur fossils in Colorado, but he was largely forced to try unsuccessfully to publish research about mountain rats while Marsh continued his series of seminal articles about Jurassic dinosaurs based on the discoveries that Lakes and others had made in the field. It was during this time that he married Edith Slater of Trinidad, Colorado. They eventually had three sons, Arthur, Harold, and Walter.

Beginning in late 1880, Lakes worked for the USGS surveying the Mosquito Mountains and Leadville District in Colorado under the direction of Samuel Emmons. Lakes prepared many of the illustrations for the USGS report about this booming silver-mining region. After his retirement from the School of Mines in 1893, Lakes lived in Denver. From 1895 to 1904, he was western editor for *Mines and Minerals*. He continued to be a prolific writer and published a number of standard works on geology in Colorado. Olive Jones's *Bibliography of Colorado Geology and Mining From Earliest Explorations to 1912* lists 259 entries written by Lakes. He retired to live with one of his sons in British Columbia after Edith's death. He died there on November 21, 1917.

The journals of Arthur Lakes include descriptions of the discoveries of enormous fossils from the Morrison formation in Colorado and Wyoming. They present a wonderful firsthand account of the discovery of dinosaurs such as the apatosaurus (brontosaurus), stegosaurus, and allosaurus. Equally important are the vivid descriptions of life as a pioneer paleontologist, often told with a level of detail and humor lacking in many scientific reports. Anyone who has ever had the misfortune to be sprayed by a skunk or injured at a dig site can relate to Lakes's descriptions. The accounts of the attempted wrecking of the Union Pacific train by desperadoes who were subsequently lynched or of the Ute War that occurred less than two hundred miles away during the fall of 1879 give a sense of wildness and isolation as great as that experienced by Roy Chapman Andrews during his Mongolian adventures.

The journals provide unequal coverage of some key events. The digging for fossils at Cañon City during the summer of 1877 and further digging at Morrison in 1878 are totally absent. The length of the descriptions of nonpaleontological matters, be they the local fauna and flora, his coworkers, and incidents in camp, may be maddening to the historian who would greatly love to learn more from this witness to some of the most important paleontological discoveries. Balanced against the lacunae are the charm of Lakes's descriptions, their delightful additions to the history of the Old West, and the amazement and excitement with which he records the discoveries that are preserved here.

We find bones of herbivores and carnivores often mingled with teeth of both species The carnivores are sabre shaped thus 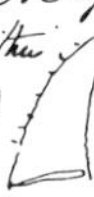 herbivores of a blunter and more spoon shaped character thus

The Lake at Sunset

Walked out by the shore of the lake at sunset there was a musk rat swimming in the rushes and a couple of ducks flew close in under the bank unconscious of my presence the female quacking lustily I could have dropped a stone on them. It was a wild scene with the glare of the sky reflected on the lake so far away and solitary on the desolate wyoming prairie

Aug 30

Discovered portions of a large sacrum the mass of vertebra three or four of them joined together to this mass the tail is attached. Reed unfortunately tapped an alkali spring which poured into one of our holes. We found some beautiful vertebra, black as ebony and shining like polished iron Some of the small ones look like reels of cotton

Took a turn with the men at the pump handle of the car saw herd of antelope from the car. Reed started in pursuit. The herd started. Reed lay down on the prairie and commenced a bombardment firing shot after shot at the flying herd till they got out of range at last one dropped out from the phalanx

1878

# Journal of Travel and Exploration

## History of the Discovery of Dinosaurs at Morrison, Colorado

One Monday in the month of March [March 26, 1877] I rode down Bear Creek Canyon from Bergen Park after my Sunday's duties in the mountains.[1] Captain Beckwith, who was staying at the hotel (Evergreen) asked me to spend a few days with him for the purpose of completing some geological examinations of the neighborhood which we had entered upon a few days before.[2]

At Morrison is a remarkably fine development of the stratified rocks of the Triassic, Jurassic, and Cretaceous periods.[3] The sandstone lying uplifted along the base of the mountains forming long parallel ridges or "hogbacks," as they are called, with [a] low valley between them, whilst Bear Creek issuing from the precipitous walls of the granitic mountains through Bear Creek Canyon cuts through the whole series of rocks and gives an admirable section. The upturned strata following one upon the other remind one of the great geological volume being half-opened and nothing remaining for the geologist except to count the leaves, measure their thickness, and examine their contents. It was this section that I was anxious to copy and measure the thickness of the rocks of which it was composed as well as examine the material, fossils, etc. helping to form them.

Accordingly tape line in hand and enlisting the help of Mr. Young,[4] the superintendent of the stone and lime quarries, and doctor_____,[5] we sallied out amongst the red masses of stratified sandstone rising tier upon tier upon the flanks of the mountains and after exploring the caves and crevices which abound amongst them, as well as jotting down holes of the granite pebbles and coarse sand of which the first series those of the Triassic are formed. The wavy and irregular character of the bedding and both of which indicate the deposition of the rock through the medium of violent water either that of torrents bringing down coarse pebbles into an inland salt lake lapping the mountains or an estuary of the sea or possibly a shore line of angry waves beating against the foot of the same Rockies at the time when they could hardly be called mountains at all being only just raised above the surface: Noting down all these points and then measuring the thickness of these massive rocks which we estimated at about. . . .

We returned to the hotel to dinner reserving the less interesting Jurassic composed of variegated red and white limestone and sandy shales and red marl. together with the high Cretaceous hogback that rises like a mighty ocean wave capped with a crest of rugged sandstone for six hundred feet above the village for the following morning. The afternoon was spent in sketching the outlines of the section and in entering notes of the details gleaned in the mornings walk.

The following morning the weather being very fine Captain B. and I started alone to a point about three miles north of Morrison where the rocks composing the Cretaceous hogback are well exposed rising tier upon tier of yellow and brown and gray sandstones like so many battlements of a fortress defending the slope of the hill. Here we began our task of measuring the thickness of the strata by ascending the hogback over the basset edges [outcroppings of rock at the edge of a stratum] of the upturned strata which at this point stand at an angle of about 50 to 60°.

## Discovery of Fossil Bones

We had not climbed very high when Captain B. who was holding the end of the tapeline in front called to me to come and examine what appeared to be the fossil compression or cast of a branch of a tree upon a loose slab of sandstone. Knowing that we had found trunks and branches of petrified trees in the rocks capping the summit of the hogback, I was not surprised at this discovery. But on looking at the impression I saw at once that it was too smooth a cast to be left by any tree and further on one end were little patches of a purplish hue which I at once recognized as fragments of bone.

Here then was the cast of a very large bone belonging to some gigantic animal. But the question was where was the rest of him and where did his majesty's remains repose. We soon traced the loose slab to the parent

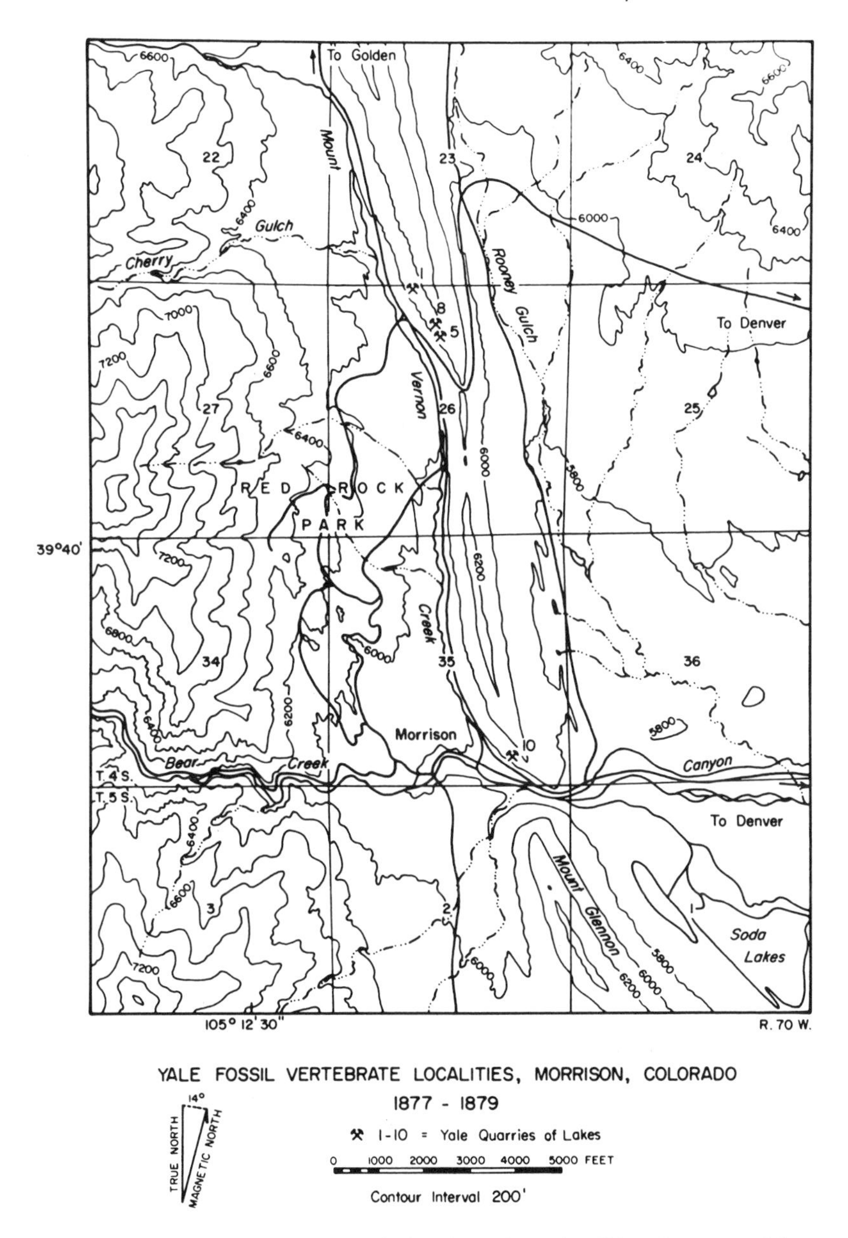


Morrison, Colorado, area and the Yale fossil vertebrate localities. Courtesy of the Yale University Press.

rock of brown sandstone from which it had slipped and as I jumped on top of the ledge there at my feet lay a monstrous vertebra[6] carved, as it were, in bas relief on a flat slab of sandstone. It was so monstrous, however, thirty three inches circumference so utterly beyond anything I had ever read or conceived possible that I could hardly believe my eyes and called to my friend Captain B. to confirm the vision. We stood for a moment without speaking gazing in astonishment at this prodigy and threw our hats in the air and hurrahed: and then began to look about us for more. Presently Cap B. cried out why this beats all! At his feet lay another huge bone resembling a Herculean warclub ten inches in diameter and about two feet long. This we supposed might be a "humerus."[7] On digging beneath where this lay under some bushes some smaller vertebrae, apparently concealed, were discovered. We concluded to defer further search and resumed our measurements and examination of the section. The hogback we found to consist generally of red free building stone quarried at the base with a series of beds of greenish grayish and salmon colored clays traversed by layers of limestone till we reached the brown sandstone on which the bones were found. Above that was another thick bed of similar clays and the rest of the hogback was formed of quartzitic sandstones of various thicknesses till we reached the cliff at the top which was composed of a fine grayish white massive sandstone with layers of finely laminated blue shale containing leaf impressions. This upper portion is called the Dakotah group[8] and the leaves of willows, aralia and other trees indicate the presence of land probably islets in the Cretaceous ocean.[9] The thickness and height of the hogback is estimated at about 6 to 700 feet or more. From the crest we looked out over the prairie towards Denver undulating like the waves of the sea.

## Further Explorations

Returning home to an early dinner we reported our discoveries and obtained the help of a blacksmith[10] and Mr. Pease's wagon[11] and I returned immediately to the spot. We hastily collected our trophies the vertebra demanding the united strength of the party to carry it down hill and deposit it in the wagon. This with several other fragments which we unearthed we laid in the wagon bed but my agony was great as our young charioteer, anxious to return home before it grew dark, drove furiously over the rough country and threatened the further dissolution and fracture of our skeleton: the value of which was but little considered by the good people of the village. The following morning[12] I had to return to my duties at the school

## Geological Time Periods

Lakes extensively used the reports prepared by the United States Geological and Geographical Survey of the Territories under the direction of Ferdinand V. Hayden as his reference source for understanding the geology of the West. The relative current approximations of names of geological periods used by Lakes in his journals are given below. In some cases, the geological period Lakes mentioned may still be used, but the period of time or the formations included have changed.

| Hayden/Lakes Term | Current Name | Stage | Time[1] |
|---|---|---|---|
| Cretaceous | | | |
| No. 6 Laramie | Laramie | Late Maastrichtian | 65–68 |
| No. 5 Fox Hills | Fox Hills | Maastrichtian | 65–74 |
| No. 4 Fort Pierre | Pierre | Campanian—Early Maastrichtian | 70–83 |
| No. 3 Niobrara | Niobrara | Coniacian—Santonian | 83–88 |
| No. 2 Fort Benton | Benton or Mowry | Turonian | 88–90 |
| No. 1 Dakotah | Dakota | Late Albian—Cenomanian | 90.5–110 |
| Jurassic | | | |
| Atlantosaurus Beds | Morrison | Kimmeridgian–Tithonian | 145.6–156 |
| Saranodon Beds | Sundance | Hettangian–Toarcian | 208–178 |

[1] In millions of years ago.

but on the next Friday [March 30] I made a careful survey alone of the vicinity discovering first a few bones imbedded in strata a few hundred yards further north and then some more in the same locality and others at the foot of the hill that had rolled down. In fact I found the ground strewed with fragments of bone but mostly of an unintelligible character whilst other remains of various sizes were observable in the sandstones. It was evident that there had been the skeleton of a monster here and that it had been a good deal broken up and doubtful how much might be left remaining to warrant the labor of excavating following the same line of rock towards Morrison. At about a quarter of a mile from this first discovery I came upon a huge bone sticking out from the rock exactly like the petrified stump of a large tree. Darkness coming on I hastened on to Morrison, and Captain Beckwith and I projected another and closer investigation for the following morning.

The weather was lovely and securing the services of Mr. Bradly, a stone mason, I started alone, Mr. Beckwith purposing to join me.

Not far above Morrison opposite the Bradly's house we detected some small fragments of bone which we afterwards learned were those of a turtle.[13] The rock however was so hard that we contented ourselves with only a small piece. For a mile or so we discovered no more remains until at a point about two miles from Morrison close to where I had found the projecting shaft of a bone on the preceding evening.[14] I came upon two enormous bones apparently the butt ends and part of the shaft of some limb bones, femur or humerus. The size of them was much greater than that of the Hercules war club and fully in proportion to the great vertebra if not belonging to a still larger animal. They were fully fifteen inches in diameter at the larger end. . . . They reminded one more of the broken columns of some old temple than anything else or perhaps a petrified pine tree stump.[15]

We now set to work upon the bone projecting from the rock and with the help of sledge and crowbar the rock was removed disclosing another huge limb bone with some ribs lying underneath it three inches in diameter.

Closer examination of the adjourning rocks revealed fresh signs of bones and on lifting off a large cap of sandstone we exposed two limb bones of a somewhat smaller animal lying in their natural position as if they had been united at the joint, apparently a humerus ulna and radius. The thickness of these bones, which lay like a plastic cast on their slab, averaged about six inches in diameter and the shaft of one bone about two feet or more in length. Several other remains of minor importance were discovered but as the sun was fast setting behind the mountains, we stopped our exploration and deferred removing our trophies till the following day.

## Snow and Carrying Bones to Town; Curiosity of People

During the night it snowed heavily and the next morning [Sunday, April 1] there was nearly a foot of snow covering the ground. Nevertheless I got the stone mason to accompany me together with the slowest of mules and a little stone cart belonging to the quarry men. The snow proved a good friend for we were able to construct a temporary sleigh and drag our bones easily and without much shaking down the cliff to the road and the wagon and we returned to the village with a load. The news of the petrified monster had spread and an eager crowd followed the wagon to Mr. Young's empty office which had been kindly lent me as a storeroom. And many hands were ready to help in unloading. But great was my mortification when a young fellow grabbed hold of the beautiful smooth polished shaft of the smaller limb bones which we had dragged down hill with the greatest care and difficulty, the bone crumbled like biscuit in his hands and my finest specimen was thus much injured. The bones were arranged on the floor of the room and a crowd as many as could be crammed into the room filled the little shanty and increased my agony tenfold by the rough way in which they handled the specimens. Many were skeptical as to their being bones at all (a skepticism we often met with afterwards) and all were astonished at the gigantic size of them speculating as to the character and magnitude of the creature whose property they were. Some after ocular demonstration had relieved them somewhat of their skepticism began to think the whole "hogback" might be the remains of some petrified monster. At last we succeeded in clearing the room and locking the door upon our visitors. Occasionally one would return bringing with him a skeptical brother whom he desired to have converted. We proceeded as soon as possible with the help of a carpenter to box the bones up and wrote to Prof. Marsh with a description and drawings of what we had found.[16]

We soon had a letter from him from Yale informing us that our discoveries were Dinosaurs of a new and gigantic species *[Titanosaurus montanus]*. The vertebra thirty inches diameter was a portion of the sacrum i.e. the vertebra to which the tail is attached and the "Hercules" war club was probably a humerus. He desired further to secure all the remains for the Yale College Museum.

Accordingly we boxed and shipped very nearly a ton's weight of bones and sent them by train from Morrison to their destination of Yale Museum New Haven Connecticut.

## Another Saurian Discovered

Captain Beckwith soon after this, wanting to go to Colorado Springs and wishing me to accompany him, nothing more was done about our "find" until the end of May when I received the means from Prof Marsh to continue and follow up my discoveries.

Accordingly I determined to camp out near the spot and spend my summer vacation at this interesting work.[17] Previous to doing so however I made another examination of the localities and picked up some fragments of bone at the bottom of the cliff close to Morrison. Tracing these up I found still larger pieces evidently belonging to some large bone. These fragments were weathered to a steel blue color. I followed these signs up the face of the hill until they ceased suddenly at a large fragment of a bone partly embedded in the soil. As I could find no signs above this point or around it, I concluded that here if anywhere the monster's skeleton must repose and commenced removing the top soil with the help of a friend. We were soon rewarded by more fragments turning up and sections of some long bones probably ribs. The further we dug the more we found and felt sure we had struck the sepulchre.[18] We hoed the top soil with great care sifting it every now and then and collecting every tiny bit of bone.

The fragments after the loose top soil was removed began to lead down to a well defined streak of dark gray clay in a bed of greenish and grayish clay underlying the brown sandstone from which further on along the ridge we had procured our specimens.

In this ferruginous [containing iron] gray streak about eight feet below the brown sandstone dipping with the angle of the dip of the hogback the bones became plentiful and evidently stratified in a regular layer.

Beneath the layer was a softer greenish clay with many round argillaceous [made of or resembling clay] concretions tinged with iron rust. In this lower layer no remains were discoverable. This layer proved very convenient enabling us to dig in it and undermine and isolate the bones without injuring them.

The deeper we dug the more packed and harder became the clay. We soon struck something which proved to be a big vertebra and close upon it another no doubt fitting on to it and some curious chambered work we at first supposed might be portions of the skull. A little to the right of this was a longish bone slightly curved lying horizontally in the strata[19] and to the left of the vertebrae similarly situated was a corresponding bone not quite so perfect. These were probably a pair. More vertebrae were discovered near the first. But the deeper we penetrated the more difficult it became to get the bones out in anything like a perfect condition the small flat thin bones, of which there were several, being badly fractured.

Finding our work was likely to be a long one I determined to camp out at once close by and after securing the services of two of my former pupils, George Cannon and Thomas Elliott to help me.[20] I went to Denver and bought a small tent and a camping outfit and about June 1st we pitched our tent on the banks of Bear Creek and our regular camp life began. The following is the diary.

June 1 Our little tent, the most diminutive possible, only eight by five and about 5 feet high called a dog tent and used by freighters and miners for shelter is pitched in a sequestered nook close to the rushing stream of Bear Creek in the corner of a crescent shaped meadow part of the ancient flood plain of the river, cut down from the surrounding valley to a depth of one hundred feet. Along the banks and covering the flat are quantities of huge granitic boulders brought here and deposited trumbly by glaciers, relics of old glacial moraines when the mountains and cañon were filled by ice. The foothills of the mountains are at this point not very lofty with smooth outlines broken here and there by rugged masses of reddish schist and granitic rocks. To see really grand scenery you must walk up the cañon whose precipitous walled chasm is indicated by a slight dip in the outline of the hills a few hundred yards west of our camp. And further if you would see the true mountains or the snowy ranges, you must climb a thousand feet or more over the intervening foothills. But for the most striking features in the scenery near our camp are formed by a group of massive red sandstone rocks rising in ruddy forms from the cool green grass of the valley and seeming as if struggling upward to climb the soft velvet slopes of the mountains upon whose flanks they lie.

## The Red Rocks

They are stratified sedimentary or water formed rocks laid out originally by violent floods of water in the Triassic period, a period teeming in some parts of the world with strange reptilian life as testified by the footprints of the Dinosaurs on the Connecticut sandstones.[21] Here they lie strata upon strata layer upon layer of massive red rock upturned from their original horizontal position by the rising of the granite mountains. Great thick leaves of the world's geological book piled one on the other their backs seamed by the action of the elements and by little streamlets formed from melting snows worn into many remarkable and weird shapes full of cavernous holes and crannies tenanted by grey foxes *Urocyon cinereoargentatus,* [red foxes] *Canis vulpus,* coyotes *Canis latrans,* owls and hawks *Falco sparverius* and the mountain rat *Neotome cinerea* of singular accumulative propensities. Indians too have used the deep crevices between the several layers as sepulchres for their dead relatives and implements used to be found in days gone by high up among the cliffs.[22]

The composition of these red rocks is a coarse cement enclosing large granite pebbles of various sizes firmly embedded in the sandstone and evidently derived as well as the other material of the sandstone from the wear of the granite mts. upon which they lie. The pebbles increase in size and quantity as you approach the point of contact with the adjacent schists, implying that heavy floods and torrents brought down the loose debris from the low mountains and deposited it by wave action as coarse sandstone along the shore of a salt lake at the base of the mountains. The strata are divided by long paralleled ravines or avenues shut in by almost perpendicular cliffs of red lurid rock two hundred feet high throwing cool shadows on the soft green grass below. Whilst here and there "growing where no life is seen" starting apparently out of the very rock itself you may see a few pine trees.

The cliffs resound with the cries of the hawks who are rearing their young in inaccessible holes on the side of the cliff whilst the little rock wren *(Salpinctes obsoletus)* pipes shrilly amongst the crannies.

Towards the upper portion the deep red of the series passes into light pink and gradually shades off into a cream color and here the Triassic group is supposed to end.

Immediately upon this lie the brick red marls and shales and variegated limestones of the supposed Jurassic forming a low hogback the softer materials exposing it to the denuding agencies. Here are some of our best lime quarries.

Next to this and shutting our view of the plains rising like a rampart against them is another hogback, the lower portion of which is supposed to belong the Upper Jurassic and the upper portion to the Dakotah group of the Lower Cretaceous periods.[23]

## Camp Life Domestic Scenes

Half way up the sloping cliff of this ridge or hogback about four hundred feet above the stream of Bear creek and close above the little village of Morrison we can see from our camp the hole out of which we have been digging our saurian bones. Up to this hole our little party of three wends its way soon after breakfast and pick and shovel are kept lively till noon, when we discuss a can of oysters and about 5PM return to camp for a more elaborate supper and our night's rest.

When we turn in for the night we are packed pretty close in our little dog tent but sleep soundly. I am generally the first to awake and look out through a little crack in the door of the tent to see if the stars are still shining or beginning to wane or whether the gray dawn is beginning to pale over the hogback. If so on the plum bush over the tent a mocking bird begins his first trial notes, a signal soon followed by a chorus of songsters in unison with the music of many waters of the stream almost beneath one pillow. "All aboard boys. The birds are singing in the treeses" is my reveille' And Tom the cook sighs deeply and gradually extricates himself from the heap of blankets and wriggles free from the close packed entangled limbs of the sleepers hauls on his boots and leaves the rest of his toilet till he gets outside the tent for there is no possible room for such domestic matters inside. We soon hear with satisfaction the sound of a crackling fire and the odor of rashers steals fragrantly into the tent and then as the sun rises over the top of the hogback and catches the peaks of the mountains steeps them with gold and lightens into a sudden glow the massive red rocks, we too crawl out to perform our toilet by the stream, water the mule, air the blankets, and set camp matters in order for the day. At last Tom cries out— Breakfast! and lays our pancakes and beans on the top of a box and we spread our blankets on some stone seats and walk in to the good fare with camper's appetites: Beau our shepherd dog is our only and much interested attendant, and the mocking bird whose nest is close by sings carol upon carol for our entertainment. Breakfast over and all about camp snug and taut the tent door carefully tied to prevent intruders in our absence, with our picks over our backs we start for our work stopping for a moment at the village post office to get mail by the early train.[24] Then we climb the hill to our hole and dig vigorously till the black end of a big bone is discovered and we all converge toward the point with great interest as to what it may turn out to be. Frequently visitors come up and watch our work with much interest.

So we worked on for some days till we had dug a trench about thirty feet long fifteen feet deep and ten feet wide. In doing this we unfortunately undermined a heavy mass of sandstone lying loose and weighing a ton or two. We for some time worked under it with a little trepidation but not conscious of the immediate danger we were in. One night it rained very heavily and this loosened the earth. The following morning we found the stone had dropped right into the hole completely blocking up our work in one corner. So we opened another trench a little to the right in the direction in which some of the bones seemed to lead and by carefully removing the top soil we were able to unpack our bones from the clay as easily as out of a meal sack. On this side we discovered several nice bones and among them a large mass which I hoped might be the skull. This mass was removed with great difficulty owing to the looseness of the clay and the rottenness of the bone. Near it was what I took to be a frontal bone and also a shoulder blade. Fragments of teeth with finely serrated edges lay near the supposed skull. The teeth were about two inches long and quite sharp. A long very slender rib was found near by and this together with other facts seemed to imply that there were the remains of more than one animal in the hole, one being smaller than the other.[25]

We concluded after our accident of the falling rock and having got out a good lot of bones to remove our work for a while to another point.[26]

June 12 So we brought down our tools and spent the day in transporting our specimens from the hole on the cliff to the yard at the back of Mr. Smith's store.[27] Tommy Elliot and I carried them to the foot of the hill and

*Digging Out Bones at Morrison, Quarry No. 10.* Watercolor by Arthur Lakes. Courtesy of the Peabody Museum of Natural History, Yale University.

then he wheeled them over in a wheel barrow to a shed. The weather was intensely hot and at noon we were glad to repast oneselves with an oyster lunch. From the store we took them to the depot labeled and packed them and sent them off.[28] George C. went mule back to Denver. I attended the funeral of a young stranger named Falais who died at the hotel in Morrison. He had no relatives to attend the funeral. A little party of villagers accompanied the coffin to the grave. A most romantic spot. A little cemetery close to the great red rocks at the foot of the mts. commanding a beautiful view of the valley. Here the little group gathered around the grave. The Methodist minister [Rev. W. R. Roby] read the impressive episcopal burial service. A few of the wild flowers of the region were thrown upon the coffin and we all stayed till the grave was filled. Mr. R. then made some appropriate remarks and the party walked sadly home not without a thought of the lonely funeral of a stranger so far from home in these mountains. The weather was very hot but the cool grass was blooming with white *Oenotheras* [evening primrose] and *Astragalus* [milkvetch], the flowers which were thrown into the stranger's grave.

June 13 We started in good season for a point further north. George and I went round to the other side of the hog back to borrow a sledge from some men who were working on a lime ledge (Cretaceous No. 3) [Niobrara]. They showed us some large *Inocerami* shells and expressed some interest and surprise when I told them they were working upon and digging up the bottom of an old sea and that an ocean had once been here.

Shouldering our hammers we walked along the base of the hogback for about one quarter of a mile and then began to ascend it looking for fragments of bone fallen from the cliff above. Some of these fragments we found loose but others were in brown sandstone spotted with greenish white concretions of clay. We found some bones resembling ribs of a purplish color. Others were of an indefinite form. We soon traced these up to the ledge or parent rock. A stratum continuous with those from which we procured the other saurians, although of this I am not certain and am suspicious it belongs to a lower stratum. At this point we found numerous small broken bones of some small animal since determined to be turtles.[29] We hunted over some tons of rock but with little success. The bulk of the animal whatever it was appeared to have rolled down in the fragments at the base of the cliff. We found a tooth amongst other fragments about one inch long and sharp. The weather was very hot and the sun's heat was powerfully reflected from the sandstone so we were glad to cool off by a bathe in the brook and returned home to camp to a supper on beans, steak, coffee, and boiled prunes.

## Getting Out Fossils at No. 5; Spines; Discovery of Other Bones; *Stegosaurus armatus*

June 14 After breakfast Jenny, our mule, was saddled and panniers [large wicker baskets attached on the sides of the animal] put on her back. We started for the scene of yesterdays labors packing down from the cliff above all the fragments of fossils and cached then at the base behind a rock. Then we rode on to the place where we dug out our first saurians in the Spring.[30] After removing several large loose masses of rock without success around a spot where we previously found a very large bone, I dispatched George Cannon to prospect around whilst T. Elliot and I continued our work.[31] We soon came upon some little cross shaped bones, apparently vertebrae. Further on was the print of a larger bone. On removing the loose masonry of rock above it our bone became larger and flattened reminding us of a collar bone. We broke open the block in which it lay and exposed twelve long black enamelled spines.[32] Tom at once exclaimed those must be his claws but I doubted this. There was a pair or two sets of these spines side by side six a piece with two small hour glass shaped bones close to them. This trophy we laid aside for closer examination and descended the hill to get out a large bone partly exposed in a huge block of sandstone fallen from above. A farmer who was passing joined us and lent his powerful arm to the sledge hammer and soon we clove the block in two but the bone was badly shattered. The cloven block however revealed many other fragments.

Whilst we were busy over this work, George C. came up struggling under the weight of a very large leg bone, "Grapes of Eschol!"[33] which he reported as a specimen of a number of bones he had discovered further on which were larger and too huge to transport. We hastened to the spot and there sure enough was another huge end of a limb bone sticking out of the clay. Proposing to further examine these on the morrow we returned to camp.[34]

June 15 Started for Golden City. I took the boys along with me to dig for fossil shells. A bank full of these we soon discovered a little way out of the hogbacks on the side of Bear Creek. They belonged to the Colorado Cretaceous No.3 and were a species of clam, *Inoceramus problematicus*. We collected a good number standing in the water of an irrigation ditch and cutting away at the bank above from this point we turned north following the road towards Golden between the hogbacks and Green Mountain [located about two miles north of Morrison]. We explored carefully every little ravine cut by the streams through the soft Cretaceous strata and clay but without much success except a few baculites [Upper Cretaceous cephalopods] in one place. The weather was intensely hot and we were glad to leave the close confinement of the narrow ravines for the open prairie. Here again we struck another bed of shells abounding in a small bivalve called *Inactra canonensis* together with baculites and scaphites [Cretaceous cephalopods] in a greenish clay. Here I left the boys digging away and returned to Golden.

June 17 Held service in Idaho [Springs].[35] Walked up the mountains after evening service to Mr. Camp's ranch[36] a lovely road by the side of a stream. The air was cool and refreshing and the moon shining brightly.

June 18 Lovely morning and lovely view looking out over the tops of the mountains. Examined his [Camp's] mining tunnel. Met Bishop and family in Idaho [Springs] and rode with them to Golden on the Car [railroad]. Mrs. S and her children sitting out on the platform outside.[37] Visited Mr. Everett in Golden who had discovered the skull of a bison at a considerable depth in the clay near his house.[38]

## Cooking Troubles; More Bones

June 19[39] Rode to Morrison. Found George in camp in much tribulation over his first cooking experiences having endeavored to cook beans in a pot without water etc. and fry pancakes without greasing the bottom of the pan. Tom E. had been snapped up as waiter at the hotel.

June 20 Secured Mr. Shields[40] a laborer to work at $2 a day. We went to work at No. 5 experiencing great difficulty in getting out the heavy and very hard massive sandstone in which the bones were embedded. We succeeded in getting out several though nearly all broken owing to the brittle character of the fossil as well as the rock. The bones were tinged with a ferruginous color. One of them an enormous flat bone was well exposed on a massive block of sandstone weighing many hundred weight. There were several other bones ramifying through the mass which we intended to send off just as it lay to Prof Marsh.[41]

June 21 George C. and I walked in the evening to Jarvis Hall to see the closing of the school. The speeches etc. etc.[42]

## Teeth etc.; *Diplosaurus;* Arrival of Prof. Mudge

Mr. Shields in my absence had been prospecting and had opened up a good deal of work near No. 1 *Atlantosaurus [montanus]*[43] discovering several fragments in the loose soil by running trenches in various directions. A large block broken open revealed some teeth mingled with other fragments some small vertebrae with processes attached (which we afterwards found belonged to *Diplosaurus*):[44] other teeth of various sizes and a tiny jaw one half inch long.

June 28 Found letters from Prof Marsh saying that Prof Mudge of Manhattan Kansas would be here shortly to help me.[45]

June 29 Accordingly the next day as we were eating our dinner under the trees an active looking little old gentleman rode up on horseback and asked if Prof Lakes was there. He introduced himself as Prof Mudge and we were soon deep in the matter of bones and saurians and relations with Prof Marsh etc. He told me he had been for years professor of natural history at Manhattan College and had been also state geologist of Kansas. He was the discoverer of the remarkable saurians, pterodactyls and birds with teeth in the chalk of Kansas. He had spent several years under Professor Marsh's direction collecting amongst the Cretaceous strata of those regions sometimes alone and often as leader of parties sent out by Marsh to explore and collect during the summer months.

He had traversed the whole of Kansas as well as other regions in these pursuits and had many stories to tell. Amongst them on one occasion whilst wandering off alone he came upon a body of Sioux who at once rode up to him. He made himself however so agreeable to them and amused them so much by grinning with his false teeth and throwing them out beyond his gums, a feat which as they had never beheld before in any human, they begged him to do again and again to their intense delight. He also gave them to understand his party was not far off and that he was from the Fort. With these and other demonstrations he got along very well with them. Meanwhile the boys who were driving behind a hogback came in sight and their first utterance was: Indians! And they got the Professor sure. They were arming and coming to the rescue when the Indians withdrew with tokens of friendship. Prof Mudge afterwards learnt they were some of the worst Indians[46] in the country and this escape was quite lucky.[47] Their party generally consisted of the Prof and two or three young men[48] and they camped out all the summer on those burning plains many miles from any settlement living on canned meats, antelope, and jack rabbits. They used to move camp pretty often spending the day in traversing the low wave like ridges of chalk, the chalk bluffs as they are called, which follow one another with ravines between them like the waves of the sea, the hogbacks not being nearly so lofty as in Colorado and the composition of the rock being a yellowish chalk easily worked. On these rocks they frequently found the impressions of pterodactyls and birds and cut them out in blocks with great care. Also great numbers of the mosasauroid and other saurian remains were secured in Cretaceous No. 3. An immense number of pterodactyls and birds were sent from that neighborhood by his party, the larger proportion of those in the Yale College Museum which have brought that Museum into such notoriety.[49]

## A Wash Out in Camp

Amongst other adventures on another occasion the river rose suddenly and washed away their wagon, horses, and tents. They only succeeded in saving a few articles.[50] This adventure reminded us of what occurred on a smaller scale at our own camp. One day we started for our work leaving George in charge. A heavy shower came up suddenly and the stream by the side of which our kitchen and larder was fixed for convenience of nearness to the water rose suddenly into a torrent. As soon as the shower was over we hastened back to camp there to find a pretty state of things. George it seems as soon as the rain began to fall went to the tent for shelter never thinking about the stream and when the storm was over he went out to find larder and everything swept away. So there was a general search made downstream and coffee pots and pans and various culinary articles were fished up from sand bars and fallen logs against which the torrent had stranded them. We had a doleful supper on what could be found and spent the evening in attempting to dry our blankets and bedding before a blazing fire.

## Camp Life and Newspapers

Professor Mudge we found quite an acquisition to our party. He appeared to be about sixty years of age but was lithe and active as a boy and full of interesting information besides being a perfect gentleman. It was a great

Benjamin F. Mudge, probably taken in 1879. Courtesy of the Kansas State Archives.

## Benjamin Franklin Mudge

Benjamin Franklin Mudge was a pioneering scientist in Kansas during the 1860s and 1870s. Born in Orrington, Maine, in 1817, he graduated from a course of Natural Science at Wesleyan University in 1840. He studied law and in 1844 was admitted to the bar in Lynn, Massachusetts, where eight years later he was elected mayor. He moved to Kentucky as chemist for the Breckenridge Coal and Oil Company in 1860, but the next year, with the advent of the Civil War, Mudge moved to Kansas. There he served as a teacher and county superintendent of public instruction in Wyandotte County.

Mudge was appointed the first state geologist of Kansas in 1864 and the following year was elected a professor at the State Agricultural College at Manhattan, Kansas. He helped to organize the Kansas Natural History Society, which later became the Kansas Academy of Science. After a stormy controversy with the board of

treat to me to meet such a one. The following day he went to Golden and procured a tent so we had two tents beside the stream. I handed over the charge and direction of the party to him as having had more experience in these matters. A business arrangement was also made between us.

Prof Mudge was generally the first up in the morning and we could hear him whistling his favorite tune "Tis morn the lark is singing"[51] and starting the fire, a signal for the rest of us to crawl out. At noon we generally got the newspaper or *Nature* and read aloud the news whilst the steak was frying. We were all much interested in the Turko Russian War and kept regents of the college, Mudge was among several professors who were dismissed in March 1874.

That spring, Mudge was hired by O. C. Marsh and continued to work full time for him until 1878, when he also accepted a position as an instructor at the University of Kansas in Lawrence. Beginning in the early 1870s, Mudge had begun furnishing specimens of fossil flora and fauna from the exposed Cretaceous strata found in Kansas to scientists in the East, including Marsh. Fossils sent to Marsh, in 1872, included a specimen of the Cretaceous bird *Ichthyornis dispar,* which had teeth.

With the help of able assistants such as his former students Harry Brous and Samuel Wendell Williston, Mudge produced a remarkable number of fossil marine reptiles, birds, fish, and pterodactyls during the summers of 1874–76. This fieldwork added greatly to the collections Marsh was building and helped him to maintain his interest in Mesozoic fossils at a time when he was actively working as well in the collection and description of Cenozoic creatures.

Mudge's work in Colorado centered primarily at the Garden Park site at which he located a new quarry and began operations in August 1877. The following year found Mudge working both in Colorado and Kansas for Marsh as well teaching during the fall semester at the University of Kansas. He also visited Lakes and the two of them made a tour of the South Park during May 1878. In 1879 Mudge taught both semesters at the university and continued his fieldwork. On November 21, 1879, he died of a stroke in Manhattan, Kansas.

up with it lively, making a good deal of fun out of the frightful Russian names such as TergusRatsoff translated "Turn those cats off" etc. No one but those who have been in camp know the treat of even a somewhat late newspaper to men thus isolated from the world.

We still continued our excavations at *Atlantosaurus* [*montanus,* quarry no. 1]. Mr Shields getting out fragments and George and I carrying them down hill in an impromptu handbarrow made of a gunny sack strung on a couple of poles to Professor Mudge who sat under some trees at the bottom near the brook dressing the stone with his hammer and chisel.[52] We soon found the work of merely splitting the rock with chisels far too laborious and had recourse to blasting. One day we put in a blast on a large flat loose fragment of sandstone which blew the entire cap off and exposed lying like a sculpture to our great delight the almost perfect skull of a small saurian about fifteen inches in length. The skull like the other bones was nearly black and the place for the eyes was clearly defined as well as the sockets for the teeth and the snout was a singular nipper looking object reminding one almost of the nippers of a centipede with sockets for the teeth well defined. Whilst I made an accurate sketch of it, the men with Professor Mudge spent nearly a whole day in chiselling it inch by inch out of the solid sandstone and we had the satisfaction of dispatching it together with some small vertebrae and teeth found in the same block to Professor Marsh. We soon heard of its arrival. Professor Marsh was delighted it proved to be a missing link between the Saurians of the Triassic and the crocodilian genus and Marsh published a description under the name of *Diplosaurus*[53] in the *American Journal of Science.*[54] He also seemed much pleased with my sketch.[55]

*Professor Mudge Contemplating the Huge Bones while Picks Continue Excavating Nearby*. Watercolor by Arthur Lakes. Courtesy of the Peabody Museum of Natural History, Yale University.

We worked for a week or more at this spot and also opened by blasting several rocks in the same ledge a little further north where we saw signs of bones in the rocks. Several nice fragments were thus obtained but none of any great importance. At one point Professor Mudge and myself were engaged chiselling out a smooth bone from an overhanging ledge. The Professor somehow getting his head in the way of my hammer I hit him a gentle tap behind the ear and knocked him over with an exclamation of Oh! on his part and apologies on mine.[56] This was one of a few slight accidents that befell our party such as one night a rock slid down weighing about a ton on the very spot our men had been working the day previous and on another occasion the whole bank of clay gave way and fortunately knocked the man who was working under it head over heels clear of the mass only burying his feet. Another time whilst watching the men striking with the heavy sledge a fragment of stone flew up and struck me between the eyes nearly stunning me and leaving a couple of fine black eyes for me to go to Idaho [Springs] the following day to preach to my people. These with sundry slight contusions of fingers and shins provocative of some hard feeling and strong expressions made up the list of casualties for June and July.

## Move to No. 8; *Apatosaurus grandis*[57]

After a fortnight or so we abandoned the *Atlantosaurus* beds of which before leaving I took a section with George Cannon and resumed excavations at the place were George C. had found his big limb bones in the clay, Quarry No. 8. For some time our search was comparatively successful though not so very wonderful and after a time the bones began to play out when almost as we were about to abandon the spot some thin flattish bones were discovered and as we removed the clay these proved to be the sort of fringe or flat edge of a gigantic and quite perfect bone six feet in length shaped somewhat like a very long hour glass, a scapula.[58] This monster like the rest we exhumed piecemeal in sections carefully marking each little section making an outline of the bone on paper and finally an accurate sketch of the whole which we sent to Prof Marsh. After this again we found very little and after another week's work abandoned the spot. We had opened quite a quarry here finding all the bones in the greenish clay which was traversed at points by long rows of large concretions. The heat becoming very great, we extemporized an awning out of some cotton which sheltered us a good deal from the burning sun. I generally took a bath in the stream before dinner.

Quantities of beautiful flowers bloomed near camp such as *Gilia vokus*[59] but the prevailing flower was the sunflower whose habit of keeping its glaring face turned towards the East and rising sun I had not before noticed.

## Trip to South Park with Professor Scudder and Mr. Bowditch, Entomologists

August 8 About this time I received letters from Prof Sterry Hunt and Lesqueaux[60] with letters introducing Professor Scudder[61] of Cambridge the noted entomologist. Accordingly I went to meet him in Denver at the American House. He had just returned with a companion, a Mr. Bowditch[62] from a trip to the Green River and White Earth Country[63] in search of fossil insects but had so far met with but little success in that dreary and inhospitable country. He wanted me to accompany him and Mr. B in an exploration of South Park especially of the Miocene beds at Florissant. I found both Prof S and Mr. B very agreeable companions. Prof S was about forty years of age and a widower but an enthusiastic entomologist and perhaps the greatest authority in that branch of science in America. His specialty was grasshoppers. Mr. B was also a "bughunter" at present devoted to *Coleoptera* [the order that includes beetles and weevils].

At Golden I procured a couple of horses for the outfit at the rate of a

Samuel Scudder.
Courtesy of the Smithsonian Archives.

dollar a day for a fortnight's trip. Mr. S visited our museum but could detect no signs of insects on the fossil leaves only one mark apparently of a mining larva[64] but very doubtful whether it might not be merely the impress of tendril of another leaf. This was on the leaf of a *Platanus* [modern species that includes the plane tree and sycamore (Buttonwood)].

After supper the horses arrived at the hall. As my mule was at Morrison we had to practice ride and tie. It was a lovely evening and soon changed into a beautiful starlight night and the grasshoppers and locusts, crickets were in full chorus. It was wonderful to me to hear Professor Scudder detect and distinguish the notes of each species *Locustrum, Acridian, Caloptenus, Gomphocerus, Oedipode* which were grasshoppers, which were locusts *Tettix Locustic* and which crickets in a chorus that seemed to the ordinary ear wonderfully blended.[65] We reached camp about ten o'clock groped our way across the brook and under the trees to the tents.

Beau (the dog) announced our advent by a vigorous barking which brought Professor Mudge and George C crawling in dishabille from their tents. Introductions were brief "I suppose you are Professor Scudder but its so dark I can't see you." After a long chat we spread our blankets on the ground and lay down all three of us on the same blanket with our eyes staring up into the galaxy of stars.

August 9 Next morning we packed up and started. At Morrison a telegram from Marsh arrived desiring Prof M or myself to go at once to Cañon City to secure a fossil saurian lately discovered there from the hands of Professor Cope who was trying to get it.[66] I could not break my engagement with Professor Scudder and so went on our journey.

Our road lay up Turkey Creek Cañon [southwest of Denver]. We paused for a moment before entering the cañon to look back and below us on the remarkable display of upturned sedimentary rocks rising like angry waves wave upon wave of upturned sandstone of various colors pink, grey, and white towards the mountains: Triassic, Jurassic, Cretaceous and Tertiary with the ocean [of] prairie beyond. In the cañon we noticed the remarkable bending and folding of the stratified schists and the dykes of syenite [igneous rock composed mainly of alkali feldspar] crossing the road. Professor Scudder with the keen microscopic eye of an entomologist detected a fine Catocala moth exactly the color of the grey granite hiding himself under shelter of a grey lichened ledge. He was at once transferred to the fatal cyanide bottle.

By noon we reached a farm house and dined on ham, eggs, and moun-

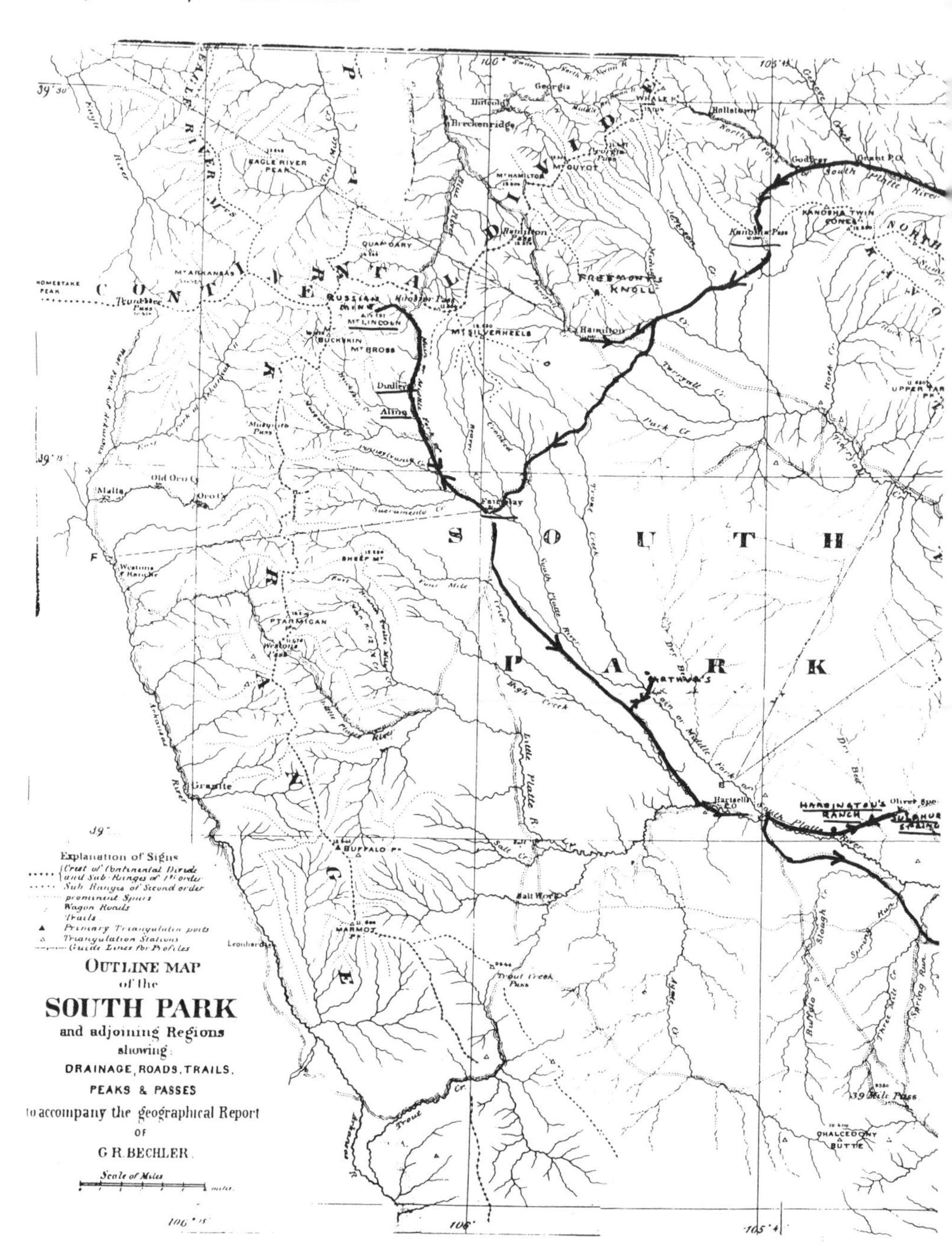


Arthur Lakes's journey through South Park with Scudder and Bowditch, August 1877. Route reconstructed by J. S. McIntosh. Based on the preliminary map of

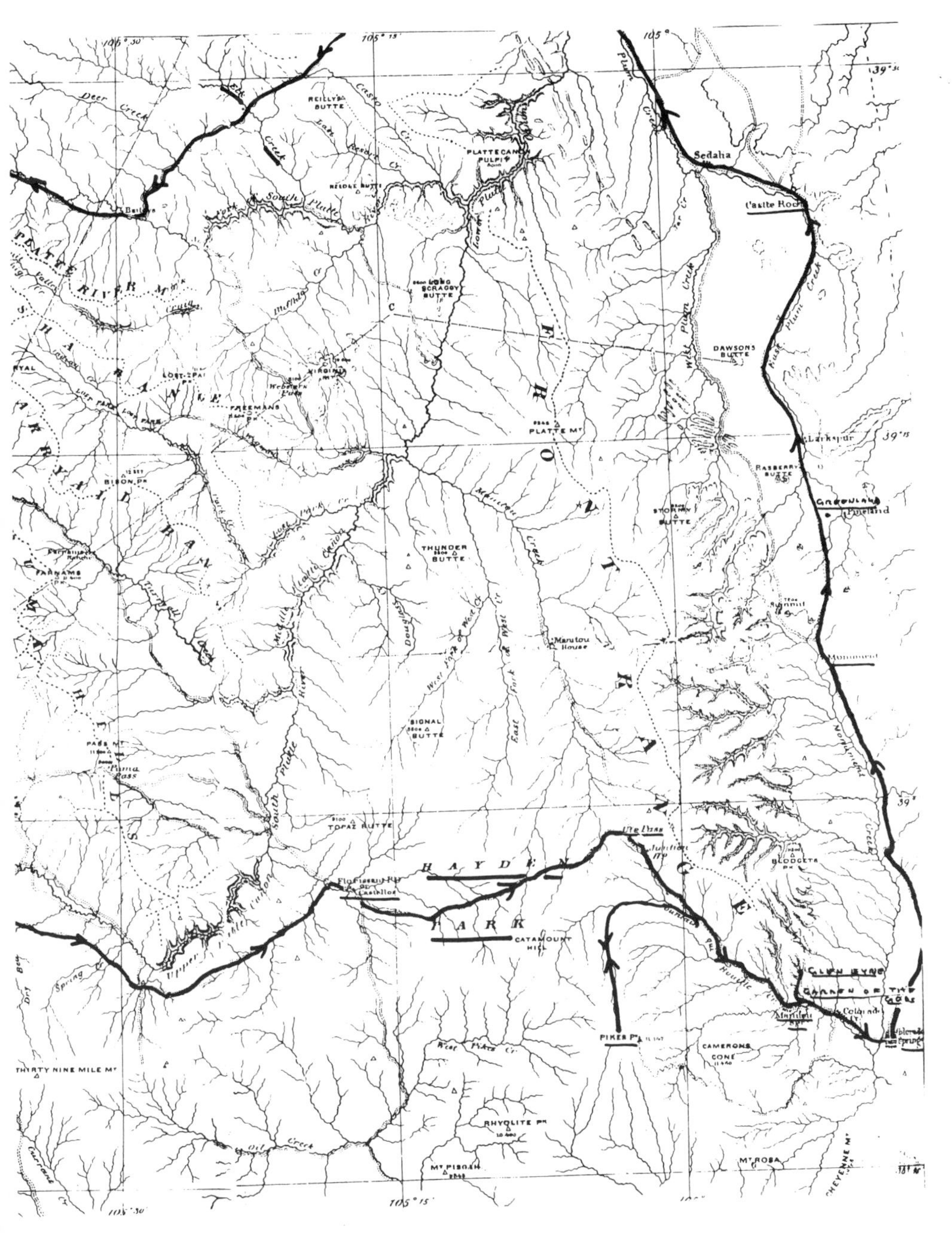


central Colorado showing region surveyed in 1873 in *Hayden Report Survey of Colorado Territories* (1873).

tain raspberries for which we paid fifty cents. The slopes of the mountains were covered with raspberry bushes of full fruit and many beautiful flowers frequented by numberless butterflies and other insects often causing Professor Scudder and Mr. Bowditch to dismount. The commonest species were Camberwell beauties,[67] skippers, Blues, Fritilaries, tortoise-shells, and commas. About 4PM we reached the Junction House hotel were Mr. Brooks pressed us to stay over night and join in a game of croquet with the ladies[68] but we were very anxious to reach Elk Creek before night. So we declined notwithstanding the attractions of the ladies, the croquet and an harmonium whose notes sounded pleasantly from the hotel.

By evening we reached S's on Elk Creek[69] and Professor Scudder and myself went for a wash in the stream whilst supper was getting ready and soon after we retired to our comfortable beds only to be kept awake by the ceaseless barking of some half dozen dogs at the coyotes.

August 10 After breakfast we had a fine view of the bright red porphyry [rock containing large crystals in a fine-grained igneous matrix] cliffs on the banks of the stream rising to a height of a thousand feet.

Took a track called the cut off which led us through meadows and forest with occasional vignettes of snowy ranges. A few grasshoppers and other insects were collected by the party and by one o'clock reached the valley of the Platte with its high mountainous banks and crazy cliffs. We halted to dine at a batchelor's cabin with the luxury of ice and iced water preserved in an icehouse by this Sybarite batchelor. We bathed in the Platte before dinner. The water was intensely cold and pretty sharp. After dinner our entomologists set to work collecting. Professor Scudder pursuing grasshoppers with his butterfly net whilst Mr. Bowditch on his hands and knees turned over cowpats for beetles and examined rotten trees for a species of tree boring beetle called *Buprestis*.

## First View of South Park

After an hour's roving we rode on through a very pretty road by the side of the river whose banks were well timbered. Towards evening the road began to ascend towards the Kenosha house at the summit of the hill we came upon a beautiful mountain lake at an altitude of 8,000 feet.[70] Here we noticed the first bushy *Pinus aristata* [bristlecone pine] which abound on the crests of the mountains. Just before dark we reached the brow of the hill where South Park opens below you so suddenly a broad waving prairie basin shut in by grand snow capped mountains.

Professor Scudder and Bowditch were charmed and our spirits rose with the exhilarating sharp breeze blowing up from the park below and from the girdle of snow-capped mountains overwhich as usual many storms were gathering.

Tired and hungry we rode down into the basin to Mr. Brubaker's hotel or as Prof Scudder insisted upon calling it "Mr. Bull whackers"[71] hotel just at the edge of the park. Here they gave us a good supper and comfortable beds. The mail coach with six horses drove up in old style[72] about nine o'clock to supper on their road to Fairplay. Having turned our horses into the pasture, we likewise turned in.

August 11 Next morning whilst breakfast was preparing we climbed the hill and sat down to study the geological and topographical characteristics of the park with the help of Dr. Hayden's report and map.[73] We had a good view of the region from this point with its waves and ridges of sandstone and volcanic outcrops of granite and lava.

We left the main road and started to explore the North Eastern corner of the park where a great deal of volcanic and eruptive forces seemed displayed. We following in this direction of Hamilton[74] skirting along one of the long parallel ridges which cross the park from north to south and we found composed of course feldsparthic granite and outcrops of a dark trachyte [a light-colored igneous rock consisting essentially of alkalic feldspar].

Prof Scudder and Bowditch collected some very bigbellied *locusterious* or as they are called "Salt Lake crickets" [also known as the "Mormon Cricket," *Anabrus simplex*] so fat and stout and unwieldy that they invariably tumble head over heels when they attempted a jump. The Rocky Mountain locust or *Caloptenus spretus* was in great numbers in the marsh several pale yellow butterflies of the order [genus] of *Colias* [commonly called sulphurs] chased one another across the meadows. The air was delicious and the scene quite enchanting. Towards noon we reached the edge of a marsh separating us from Fremont's Mound, a little hogback fortified by honeycombs of loose stone rifle pits attributed to Fremont but much more probably one of the Indian forts so common in the park. The story is that F fortified himself on this hill and stood looking on at the conflict of two hostile bands of Indians.[75]

We floundered through the marsh up to the bellies of our horses the whole bog rolling like a sheet under our horses' hoofs. At last we reached the mound which is a dyke of eruptive trachytic probably about one hundred feet high. On the top was a rude fortification of rifle pits made of loose stones, a capital situation for a fort as on one side it was protected by the marsh and on the other no foe could approach on horseback without exposing himself for a mile or more to the fire of the fort. No tree or rock in the vicinity allowed a shelter or chance of sudden surprise.

After lunch Prof S. collected a number of the curious spiders which build their webs in the grass weaving the tops of the grass like a craal [enclosure for cattle] or lobster pot on top of which it forms its nest and lays its eggs. The spider, apparently a geometra,[76] was not very large but marked on the back with a distinct cross. Whilst they collected I made a sketch of the fort.

## Drenching Rain; Tertiary Strata; Fairplay; Moraines

Heavy clouds began to rise all around the edges of the park basin and soon the bitterly cold rain descended in torrents upon us drenching us to the skin as we had nothing but our camping shirts on. This continued for some hours. On the banks of the river Tarryall we found a bed of sandstone underlying the glacial drift and auriferous [gold-bearing] gravel containing obscure fragments of leaves. The stratum was a soft yellowish sandstone evidently Tertiary and of a yellow color but we had no time to explore it but passed on to the Buttes near Lechner capped with trachyte and thence to the hogbacks of the Dakotah and Cretaceous flanking Trout Creek thence across the red beds to Fairplay:[77] as we neared this mountain town the clouds began to lift revealing the dark skirts of the mountains but concealing their summits and upper portions.

We rode rapidly through the town well stared at by the inhabitants and took the road to Alma passing on our way the remarkable morainal deposits which are much crowded in at the point where the Platte cuts through them above Fairplay. In other places they form mere rounded mounds on the slopes of the mountains apparently as lateral moraines, which you can follow some distance up the deep ravines of whose material they have been formed. These ravines frequently head up into an amphitheater of which there is a remarkable one on Four Mile Creek called the Horseshoe. Towards sunset we rode into Alma[78] passing a square board nailed to a tree.

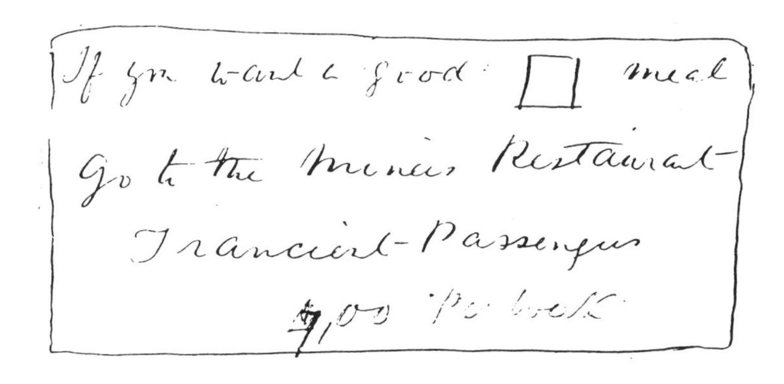


We reached Alma just as it grew dark and gladly drove our drenched horses into Mr. Murrow's stable and betook our drenched selves to the St. Nicholas hotel [and] to supper at Mr. Murrows. I heard of the arrival of Cousin P[ellar] at Arthur's.[79]

August 12 A pleasant quiet morning rested and visited Mr. Williams and wife. Christened their baby.[80]

## Ascent of Mt. Lincoln

August 13 Monday morning was very fine as we rode up through the streets of Alma and to Dudley to the foot of the mountain.[81] We passed on the way by the side of the ravine whose bottom is occupied by a branch of the Platte [middle fork] some very fine morainal deposits crowded together in rounded banks on the slopes of the mountains and banks of the stream. One of these moraines is utilized as a cemetery by the people of the neighborhood. The boulders composing these moraines are all derived from the adjacent mountains and are composed of mica schist, quartzitic gray, pink, and dark bluish black magnesian limestone, gray and variegated porphyry with long oblong crystals of a mineral resembling feldspar. All these rocks are to be found in place in the mountains around.

The vast excavations of the amphitheaters and cañons afforded abundant material for these heaps of detritus and rubbish originally cloven out of the mountains and distributed by the glacier in banks and moraines and subsequently modified by streams and floods.

A very interesting study in ascending these mountains is that of the various zones of life and fauna and flora. These are not easily determined or defined or separated as there is no sudden line of demarcation but you pass so insensibly from one zone into another that you are only conscious of a gradual change where you perceive that the general appearance of the trees, flowers, and plants is different to what it was below you. And this change is so gradual owing to the fact that though instantly or suddenly new flowers begin to appear yet the old ones we have passed below pass up with them some continuing throughout nearly to the top others dying out and disappearing and replaced by new forms. Those that have held their own from quite low down become dwarfed in character as they ascend. This gradual transition of zone into zone reminds one of the gradual passage of geological periods into one another. At the base of the mountain we have the subalpine flora which are most luxuriant and brilliant especially along the sides of the charming little brooks which descend the mountains and cut through the moss. The following list are amongst the most characteristic beginning from the base up to timberline[82]

| PLANTS AND INSECTS | ALTITUDE |
|---|---|
| Fireweed | [ca. 10,000 feet] |
| Yellow—*umbelliferae*[83] | |
| *Potentilla* | |
| *Achilliea* [yarrow] | |
| Aster | |
| Wild strawberry | |
| *Castilleja* [Indian paintbrush] | |
| *Senecio* (auria) [goldenrod] | |
| Mountain clover | 11,000 feet |
| *Rosa blanda* (wild pink rose) | |
| *Melitaeinae* butterfly [probably *Phyciodes campestris*] | |
| Harebell | |
| *Mertensia* [mountain bluebell] | |
| and some *carexes* (sedges) | |

ABOVE TIMBERLINE

Sedges *(carexes)*
White bunch top[84]
Purplish red robins with striped salmon colored bills
Purple alpine star (moss)
Dwarf Potentilla
*Carices* [sedges]

*Ribes* (gooseberry) larva
*grapta formi* on it[85]
Mertensia — 12,000 feet
Parnassius[86]
*Artemisia* [sagebrush]
Dwarfed Caloctroitus with large flowers
*Cirsium Drummondsii*[87] — 12,500 feet
Aquilegia (Rocky Mountain columbine)[88]
pale flowers
*Claytonia Virginica*[89] — 13,000 feet
Arctic mosses white and blue
*Polemonium confertum*[90] — 14,000 feet
we found at the tip top of the mountain

TREES [not in sequence by altitude]
*Pinus Aristata* [bristlecone pine]
Common spruce [Englemann spruce]
Silver spruce [Colorado blue spruce]
*Juniperus communus* [common juniper]

Above timberline a species of willow[91] continues to a considerable height.

The inclination of the trees down the mountain towards the East slope implies that the northwest is the most prevalent wind.

Many of the subalpine plants continue up to timberline whilst others pass up into the higher zones.

Amongst this particular class we may mention the *Castilleja* which is of a flaming scarlet on the plains and even in South Park passes into a variety of a beautiful salmon color at the base of the mountain and mixed with it we find another variety of a beautiful cream color which increases in abundance as we ascend and keeps on up to at least 13,000 feet in a dwarfed form whilst we lose the salmon colored variety at about 12,500.[92] The *Mertensia* also holds its way up to a considerable height above timberline and in proportion as it diminishes into a dwarfed form the color of the flowers intensifies in blue.

At 11,000 feet we found the first mountain clover and followed it up to about 13,000. The *Campanula langsdorffiana* or harebell becomes very luxuriant above timberline and what it may loose in length of stem it gains like other plants in luxuriance size and length of flower.

The *Aster, Potentilla, Achillea,* and a species of aster continue up to timberline and for several hundred feet above it.

Of the subalpine trees there appear to be three or four varieties continuing up to timberline.

The predominant tree seems to be *Pinus aristata* but struggling up with him towards the higher regions are some varieties of spruce such as the *Abies menziezii* [Douglas fir *(Pseudostuga menziesii)*].

The *Juniperus communis* continues up to and beyond timberline as a low bush and a species of willow forms with the Mexican greasewood[93] the most prominent and only brush covering the slopes with the stunted spruce for some feet above timberline the willow continues to about 13,000 feet. The characteristic changes the trees undergo in their external appearance in their transition upwards is very marked, the butts become enormous and the roots equally huge grip the ground for many jaws around the base like huge knotted serpents. The trees very suddenly taper. The lateral branches are short and the branchlets very full of twigs, the foliage very stiff and bottle bristling and every appliance is used by nature to adapt them to the terrific storms to which they are exposed. After we pass timberline they crawl along the ground like shrubs whilst a long row of bleaching skeletons tells of the forefront of the fight who succumbed before the blast.

Our entomologists were delighted with the immense number and variety of species of insects they found above timberline many of which were entirely new to science. At the top of the mountain we found a great number of black and very active spiders taking shelter underneath the loose stones.[94] These stones also sepulchred millions of Rocky Mountain locusts who had perished in some previous season in trying to cross the range their carcasses lying nearly an inch deep under the stones where they had sought protection from the elements and where the cold and snow had preserved like mummies their corpses on the snow bank at the summit. Grapta.[95] Prof Scudder discovered also the bodies of a number of small insects of different kinds which had been blown up there by wind storms. The sole living insect inhabitant of these top most rocks seems to be the aforementioned black spider and I have sometimes seen the great unwieldy "Locusterium" or Salt Lake cricket very near the summit but he is only a visitor at best.

We stayed for a short time near the flagstaff enjoying the unrivalled view and then hastened down to the Russia mine to dine having been invited to do so by the proprietor Dr. Dougan.[96] After dinner we visited the mine with its wonderful display of frostwork which we have described elsewhere.[97] Story of Mountain Rats *Neotoma cineria*.[98]

On our way down the hill we found a little brook cutting through the moss on its banks were abundance of the purple *Primula parryii* [Parry primrose] growing luxuriantly at about 13,000. At last I succeeded in drawing off the enthusiastic entomologists from their insect treasures of the mountain and we returned home through the deserted mining camp of Quartzville[99] gathering quantities of subalpine flowers for our hostess in Alma.

## Buckskin Cañon

The following day we rode up Buckskin Cañon which runs up on the west of the town of Alma and is occupied in the middle by the small stream. The cañon is about one half mile wide and ascends or heads up gradually to a magnificent amphitheater or *cul de sac* the starting place no doubt of the glacier and other forces to whose agency the cañon is due. This amphitheater forms the division between mountains Bross and Lincoln and you look down into [it] from the top of these mountains. At the bottom of the amphitheater are enormous banks of debris lying in ridges and moraines some of these are turfed over. On tops of these banks are striated and polished red boulders indicating glacial action. These moraines are cut by streamlets on whose mossy banks the beautiful *Primula parryii* grow luxuriantly together with a species of cream colored *Ranunculus* [buttercup].

Near the head of the amphitheater is a lake covering about two acres.[100] This is one of the sources of the River Platte, its waters look emerald green from above but rather black on nearer approach. High up in the cliff above one forming the center of an amphitheater the other filling a rocky basin from both of these noisy little streams fall in cascades over the rocks into the gorge below and unite with the fork of the [South] Platte. There are some profound amphitheaters high up on the western side of the of the cañon without in some cases the V like ravine leading up to them and forming their outlet. They seem almost like pits. We had no time to examine any of them but from the streams issuing from them they doubtless contain small lakes in their center. On the east side the cañon is red with cliffs stained with hematite red and yellow with much debris at the base. The centre of the cañon is occupied by Buckskin Creek overgrown by willows. Many flowers of wonderful brilliancy grew on the bank, amongst them great quantities of the *Aquilegia* [columbine] whose flowers were particularly luxuriant and large but of a pale hue, almost white in some cases, the pale white and the darker tint growing on the same plant. The portions of the cañon intervening between the head of the gorge and its outlet seemed comparatively free from any great amount of glacial debris. The lateral moraines may have continued along the slopes concealed by timber but it is not till you come towards the outlet about three miles from the amphitheater and source of action that the moraines appear prominently and then crowded together and heaped in chaotic confusion. After this we come out on the flood plain flat or park of glacier. A phenomenon I have frequently noticed and consider the glacial cañons under these divisions:

First, an amphitheater at the head of the gorge with moraines.

Second, an opening of the cañon with lateral moraines and glaciated boulders rather high upon the slopes of the cliffs.

Three, at a narrow point of the cañon a general huddling together and crowding in of moraines which at these points may have dammed up a lake till the stream cut through them.

The flat referred to was probably another glacial lake. The lateral moraines keep along the edges of the flat and continue down the course of the hills till Buckskin Creek opens into the broader valley of the fork of the Platte. This valley is another great flat or glacial lake bottom extending from the Hoosier Pass[101] and the base of Mt. Lincoln below which is its head or cul de sac with moraines on either side of the stream reaching down to the turning off into Fairplay where there is again a huddling together of moraines which must have dammed back the lake which filled the Platte Valley.

Into this lacustrine valley [one in which a lake formerly existed] all the lateral moraines of Mosquito, Fourmile, and Buckskin gulches seem to have emptied thus contributing to the broad flat of the Platte. The moraines are huddled in together near Fairplay and the Platte cut through them into the auriferous gravel which is no doubt the ground moraine or detritus of the glacier. The course of the stream turns off at an angle towards the south at Fairplay into another broad flat plain more than a mile wide which must have been one of those flood plains half river half lake formed probably at the close of the glacial epoch by the melting of the glaciers or else the broad valley of the now diminished Platte. The whole plain is covered with water worn pebbles to some depth. This valley or flood plain lies between the red rocks of the Trias on one side and the uplifted ridge of the Dakotah group on the other. The Jurassic seems wanting or else has been superficially eroded away. The ridge of the Dakotah is well defined being generally traceable by the growth of pines on its summit. Geological formations often decide the growth of timber and herbage. The hogback is

divided into a series of benches formed by different hardnesses
does not in general attain the height of that along the Plains and
border.[102] The elements composing it are however somewha
that on the Plains as will appear by the following section.

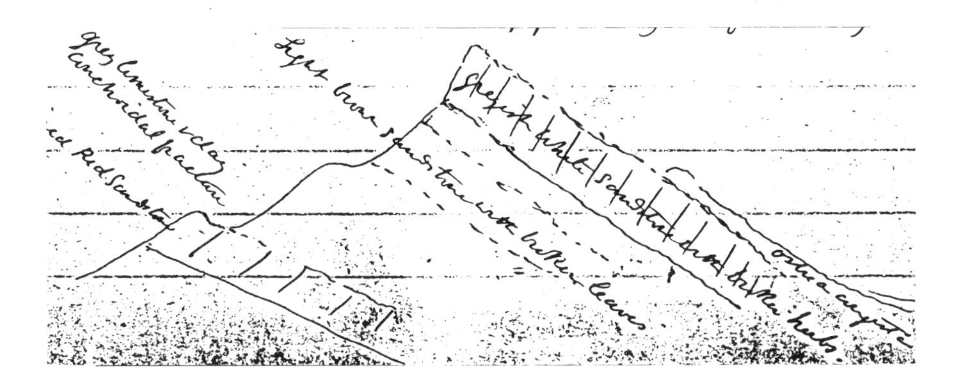

The portions containing our Dinosaur remains i.e. the *Atlantosaurus* beds are either not represented at all or are generally covered with grass. The latter I found to be the case for many miles. The lower red beds are tolerably well represented and persisted and so is the top characteristic ridge of light grey sandstone.

## Mr. Arthur's Ranch; Prairie Dogs and Wild Fowl

On reaching Mr. Arthur's ranch we found my cousin Peller and a college friend and pupil Mr. Carve there on a visit from England. The A's had enlarged their house and together with children and the farm men and visitors we sat down over twenty of us to table, Mrs. Arthur assisted by a young lady in the house having to cook for us all.

After breakfast we climbed the volcanic fold east of the house and spent some time endeavoring to follow the strata according to Dr. Hayden's report and maps. In the evening Mr. Scudder, being troubled with asthma, preferred sleeping in the hay in the barn. Mr. A and C rode to Alma. I rode with them part of the way whilst Mr. S and B spent the afternoon in collecting grasshoppers and beetles.

The ranch struck us with its usual wildness and isolation. The little Killdeer Plovers running along the prairie close up to the door of the house or screaming plaintively along the margin of the ponds which were as usual covered with flocks of waterfowl, ducks, and gulls (Heering gull? *Larus argentatus*).[103] The whole field and prairie resounding with the teasing bark of innumerable prairie dogs *Cynomys [ludovicianus]* who stood erect like posts on the top of their borrows or ran to and fro to visit and gossip with a neighbor. They have grown very tame around the house and stand on their holes a few feet from the path barking incessantly. One which stood erect in his hole with paws drooping in the most comically servile attitude allowed me to sketch him on the spot. He remaining in this attitude for full five minutes.

Next morning soon after breakfast we bid the Arthurs adieu and started again on our journey through towards Colorado Springs following the course of the River Platte. Our road lay along the Dakotah hogback between it and the mountains. For some part of the way our journey was pleasant enough till we reached the borders of a marsh where we passed through an ordeal of tiny black flies that bit our head and ears unmercifully. Occasionally we had a variety in favor of swarms of winged black ants *Formica* that pattered down on us like drops of rain and whenever disturbed the whole swarm inflicted a simultaneous sting as if at some signal or by preconceived arrangement like so many red hot needles penetrating the skin. This was exploring under difficulties and we were thankful at noon to reach Hartzell's Ranch[104] and stop for dinner at a point where the Platte cuts through the Cretaceous hogback and where there is a hotspring.

## Dinner at Hartzell's; Eagle and Hawk

Whilst waiting for dinner we noticed a fine golden eagle *Aquila chrysaetos* hovering over the opposite bank pouncing at last and striking with his talons at a prairie dog. He missed his aim however and the dog escaped whilst a large sharp shinned hawk? disapproving of such poaching on his preserves pursued and drove his fellow bird of prey from the neighborhood.

At dinner two damsels waited on us relieving guard in the laborious duty of fanning off the plague of swarms of house flies with an extempore and original punkah made of loose ends of papers tied to a stick. We paid 50 cents a piece for our dinner and resumed our march by the side of the Platte. Our road was now no longer running parallel with the ridges of strata which traverse the park like long waves but right across them. These ridges are usually separated from one another by soft meadow like valleys with here and there a few low outcrops or terraces capped generally with dark volcanic rock. The first trough between the ridges after leaving Hartzell's seems paved with Tertiary sandstone if we may judge from an outcrop of shales we found on the bank of Trout Creek where it opens into the Platte: The shales here which are finely laminated dark slatey color show on splitting great quantities of chopped vegetable and carbonaceous matter with some tolerably perfect leaves also some indistinct signs of insects for which Prof S and myself hunted industriously. On the banks of the creek we came suddenly upon two wild geese *Branta canadensis*, common wild geese strutting along as unconscious of our near presence as we were of theirs. They no doubt build here.

Leaving this point only a few yards brought us to the foot of a high bluff 300 to 500 feet high and we were surprised to find that on the western flank the stratum of brown sandstone again resumed its easterly dip showing that a fold must exist at this point. A point which seemed probable to judge from the amount of trachytic and granitic boulders scattered over the mound. I climbed this bluff alone on my agile Jenny, the horsemen declining the deep ascent. I found on the summit one of Dr. Hayden's stations marked by heaps of stones.[105] North of us separated by meadowland was another bulky mound split up by little cañons on the S end appeared sedimentary strata but we had no time to visit. Crossing another prairie meadow about a mile south we rode towards a butte on the N side of the road close to Mr Harrington's house.[106] This proved to be capped with lava and trachyte some of it of a handsome greenish color. On the Eastern slope in some dark black shales apparently much metamorphosed I found some small well defined bivalves resembling *Cyrena* or *Corbicula* or the Cretaceous *Mactra canonensis*.[107]

## Sulphur Springs; a Sulphureous Hotel

Thence we rode to the Sulphur Springs.[108] The evening was cool and our trip most pleasant; still we were glad to reach the Sulphur Springs at 8:30.

These springs are situated at the base of a volcanic mountain which no doubt has something to do with the origin of the springs. There is a singular smell around the place reminding one not a little of the smell of a sea beach. The water tastes and smells particularly like rotten eggs. Judging from the sickly appearance of the proprietors of the springs and all connected with it we should judge that they had partaken largely of the waters and that they had not particularly agreed with them.[109] There was no hay for our horses. We had to wash in sulphur water and everything about the place partook of this sulphureous nature. It would not have been difficult to imagine ourselves in close proximity to the mythical infernal regions. There were sulphur springs in the yard and sulphur water to drink and wash in in the house and the vegetables were cooked in sulphur. We remonstrated and growled somewhat at our host but he assured us that after a time we should get to like it, on the principle I suppose of which the eels get to like being skinned. We were up early next morning and glad to quit this sulphureous region.

We were up early and started in the sweet cool morning to explore the region of the park within a circuit of five or six miles of these sulphur springs. Our first point was towards a high mound capped apparently with a castle of trachytic rock fifty or sixty feet high SW of the sulphur spring. On examining an exposure near the base of under a mound apparently capped with granite, I found a layer of black concretionary limestone containing to my surprise well defined Cretaceous fossils baculites and scaphites of No. 4 Cretaceous [Fort Pierre]. I was surprised for I had supposed all this portion of the park to have been of Tertiary origin.[110] The only way in which I could account for the Cretaceous strata being that the form of the park is crescent shape, thus the center was probably Tertiary and we had come in upon the ends of the horn.

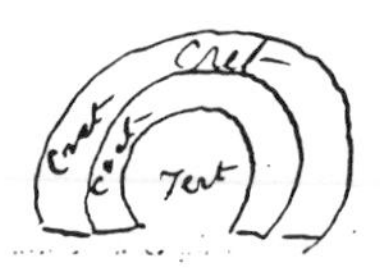


Further on towards the southwest we crossed the Platte climbed the opposite bank whose slope was much prospected by prairie dogs showing a yellowish soil underneath the top of the bluff was covered with drift and quartzitic and lava pebbles. Keeping towards the SW we came upon a prairie dog town with a badger perched upon a hillock. As we rode towards him he slowly retreated into his hole closely pursued by Beau. On reaching his hole Mr. Badger *Taxidea americana* decided to defend his home and castle rather than retreat inside it and a regular set to followed between him and Beau, the badger hissing like a cat and showing his formidable teeth which did not deter Beau from tackling him. I was much afraid my dog would get hurt and so advanced to the rescue with my geological hammer. Beau thus reinforced made a furious lunge at the badger just in time to receive the blow I had aimed at the beast on the side of his jaw and dropping the fight he fled howling over the prairie no doubt attributing the blow more to the badger than to me. I dispatched the brave brute feeling however like a murderer in killing a harmless animal so brave a defender of his home and property; but I wanted his skin. Close by we came upon an open marshy place at the foot of a mound capped with greyish sandstone at the base of which were calcareous shales with fragments of some large shells probably Cretaceous No. 2 [Fort Benton] or 3 [Niobrara]. Just here another sulphur spring bubbles up. Further on to the west we came upon *Inoceramus deformis*[111] then high bluffs capped with trachytic riding towards the North we passed several mounds capped with heavy masses of dark trachytic breccia [an igneous rock composed of sharp-angled fragments cemented in a fine matrix]. On the East slope of one of these we found shales dipping vertically with an Easterly strike. In these were fragments of vegetable matter possibly Tertiary. Further on between this and the Colorado Springs road were other shales of a light color with lignite. Thence crossing the road at Harrington's we left the sulphureous regions at 2PM for Castello's.[112] As we left the outskirts of the park we came upon out crops of granite and were soon within the granite walls enclosing the basin and looking back over that lovely scene of an ancient lake now transformed into a beautiful prairie meadow transversed by long wave like ridges and dotted here and there by little dark topped islets of volcanic origin throwing long shadows over the ancient lake basin in the setting sun whilst Buffalo Peaks frowned fortress like on the other side of the basin. To the South the park is like a chopping sea of dark volcanic eruptions with long wedge shaped bluffs of agate mountains conspicuous against the horizon.

We halted for a moment on the crest of the hill overlooking the park to bid farewell to this beautiful mirage-like scene only to turn around and behold one equally sublime. For as the cañon opens we look East and find the vista filled by the magnificent mountain mass of Pikes Peak standing out alone and unrivalled above the humbler foothills the beau ideal of a mountain the lower portion bathed in deepest violet and the upper above timberline composed of a course red granite flushed by the last rays of the setting sun whilst all the valley below was bathed in evening grays.

At one place as we rounded a cliff we saw a rabbit and I was rash enough to attempt the liberty of shooting at it with my revolver from Miss Jenny's back. The effect of the report was remarkable. Jenny spun around like a dancing dervish till my head swam and nearly succeeded in unseating me saddle and all.

After some five or six miles we again struck the Platte in a beautiful park-like ravine on the turf on which were numerous tents and wagons of campers who were preparing supper. This park ravine continued for some miles till towards nightfall we descended into a depression among the hills with light grey shales cropping out of a cutting on the side of the road showing we were approaching the skirts of another old lacustrine deposit among the granite hills. By 8 o'clock we rode into our destination of Castello's hospitable ranch.

## Another Lake Basin; Castello's Ranch; Jenny Insulted

This well known ranch is a combination of a number of little houses which have grown together one after another as they were needed forming a little village of which Judge Castello is the Lord and Master. The stout old Judge, a regular fine host of the big bellied jovial old time type received us with great hospitality. Just as we were taking off our saddles in the moonlight, Jenny for the first time had an opportunity of looking around and discovered to her consternation and disgust that she had been the unconscious bearer of the dead body of the badger which unbeknown to her I had strapped onto the back of my saddle a fact which gave great dissatisfaction to my fellow travellers as it stunk fearfully and they kept a respectful distance. Jenny's fright and indignation was unbounded. She immediately sprang from my hand and dashed off into the darkness with the dead body which had got loose slinging to and fro against her like a bolus. Fortunately she dashed up into the corral where I caught her with her large eyes glaring out of their sockets in the moonlight. Two such insults in one day was more than J. could stand. On my return from the pursuit I found our hospitable host had made a roaring fire of logs in the bar room. Although the depth of August this was grateful. The logs were burning in a large open fire place the pillars and mantelpiece of which were quaintly enough formed of big blocks of the petrified stumps of the basin and the walls of Miocene shales entombing abundance of beautiful leaf impressions and fossil insects.

This little collage was the Judge's reception room and we gathered in it every evening for a chat over the fire and a smoke. For our own use we had another delightful little cottage allotted us papered with illustrated newspapers, *Harper's Weekly*'s, etc. This room we soon turned into a naturalists study littering it all over with bottles for collecting insects, fossils, geological hammers, insect nets, portfolios, and sketchbooks.

**August 18** Spent the morning skinning the badger. Prof. Scudder went over to Hill's ranch to see about some fossil insects Mrs. H. had been collecting for him.[113] The Judge took Bowditch and me up to examine a singular volcanic overflow behind the house the summit of which had been rudely fortified by the Indians and had been the scene of many a wild encounter. It was a hill about 60 to 100 feet high composed of trachyte broken into fissures and crevices apparently by the breaking of the divisions between the basaltic columns. The sides and top of the hill were profusely scattered over with loose fragments of trachyte which the Indians had piled up into rude honey comb rifle pits. The fortification was a strong one on one side but weak and exposed on the other where the fort was overlooked by a low hill. An advantage of which a band of Utes once took the better of to completely exterminate a band of Arapahoes who held the fort. One of the strong points of the position was that the deep fissures afforded excellent hiding places. The poor Arapaho squaws sought these however in vain as the Utes killed them all. A Ute telling the story says: "Ute. shoot! shoot! Squaw heap cry. Kill em every one."[114]

There were some remarkable funnel like holes in portions of the massive lava regular and smooth and circular as a funnel about a foot in diameter and penetrating in one case five or six feet and how much deeper we could not tell. The sides seemed scored by heat and I judged they were volcanic vents either of steam or water.

Prof Scudder came back at dinner time laden with spoils reporting that Mr. H had boxes upon boxes full of fine paper like shales covered with the impressions of most perfect insects of various descriptions taken from the sedimentary Miocene tablelands or mesas near their ranch.

After dinner Prof S, B and I started for Crystal Mountain a prominent peak about six miles from Castello's from the summit of which we hoped to get a good view and surveying the landscape over more especially the region of the park which we purposed exploring on our way we looked in on the hut of a man named Houghton[115] who had a fine collection of minerals from the neighborhood more especially from Cheops pyramid or Mt. Crystal [also known as Topaz Butte or Crystal Mountain]. These were large masses of orthoclase crystals: Crystals of feldspar and magnificent specimens of the far famed green amazon stone cubes two or three inches in diameter of a brilliant green color, some of the crystal single, others in large masses.[116] Mr. H was one of poor Custer's scouts. After examining his collection and buying a few specimens and some of the magnificent crystals of smoky quartz we passed on towards the pyramid. The granites by the side of the road were of a very coarse red and porphyritic character loosely compacted and easily breaking up into the coarse red gravel of the

Petrified stump at Florissant, Colorado, about 1875. Courtesy of the Myrtle Dalby Collection of the Denver Public Library.

## The Florissant Fossil Beds

Arthur Lakes's account of Samuel Scudder's exploration at the Florissant Fossil Beds describes a site that is as important for its paleoentomological and paleobotanical specimens as are Morrison and Como Bluff for their vertebrate paleontological specimens. As Scudder noted in "The Tertiary Lake—Basin at Florissant, Colorado Between South and Hayden Parks," "The insects preserved in the Florissant Basin are wonderfully numerous, this single locality having yielded in a single summer more than double the number of specimens which the famous localities at Oeningen, in Bavaria, furnished . . . in thirty years. Having visited both places I can testify to the greater prolific-

vicinity Cheops Pyramid besides in shape resemble its name sake of Egypt by being composed of huge loose blocks or stairs of granite[117] with pockets or little caverns and crevices lined with crystals of smoky quartz feldspar orthoclase and occasionally amazon stone the latter running in little veins like the feldspar. We climbed the hill and peak with some difficulty owing to the height of the huge blocks and on the top found one of Wheeler's cards of heights etc. in a little tin box protected by a pile of stones.[118] The view from the peak was all we could have desired overlooking the entire country. To the South was the huge mountain mass of Pike's Peak with a long straight wall extending from it northwards and separating the granite

ness of the Florissant beds." Scudder eventually published more than twenty-five papers related to his discoveries at Florissant. Since the 1870s, paleontologists from many institutions including Harvard, Princeton, the American Museum of Natural History, the Carnegie Museum, and the British Museum have mined this site. It is the premier North American site for Tertiary paleoentomology.

The fossil beds were created through a series of volcanic eruptions during the Miocene, thirty-five million years ago, when mudflows buried tree trunks and stumps that then became petrified. A small lake was also formed in which periodically, over many centuries, creatures were trapped in thin layers of ash when the Thirty-nine Mile Mountain volcano erupted. These layers vary in density, reaching up to forty feet in places. The lake eventually dried up and a final, thicker layer of rock fragments cemented itself together as a cap, which protected the lower layers. Among the trees fossilized were giant sequoias, one of which had a trunk 22 feet in diameter.

The dig site, the Scudder Pit, was just west of the Big Stump. In five days they collected nearly five thousand fossils, including items purchased from local collectors. Scudder kept the insect specimens for his own research. He sent the plant specimens to Leo Lesquereux and the vertebrates to Edward Drinker Cope. The collections were eventually transferred to the United States National Museum.

mountain from the foothills as well as forming the boundary of Bergen's Park to the North several rugged crags of red granite shot up from the low pine covered hills. The character of this portion of the mountains is compared to that further North of a much lowlier character. To the East at the base of Pike's Peak is a broad open prairie like park called Hayden's Park.[119] To the West the course of the Platte might be traced up from South Park to its cañon through the mountains. The scene was characteristically that of low wooded mountains or rather hills cut up by many small valleys forming the courses of numerous little streams. We returned back to supper by nine o'clock after a hard days work. As the Judge said of us you put in 23 hours out of the 24. We highly appreciated the supper of elk meat and wild raspberries both of which had been procured from the vicinity.

August 19 The following morning Prof S was up soon after daylight and we started to explore the locality known as the Petrified Stumps, celebrated for these gigantic petrified trees as well as for the beautiful fossil impressions of insects. These are found about two miles South of Castello's on a southern entrance of the lake about five miles long from two to three in width with many little bays and inlets on either side running up amongst the timber and drained by the branches of Twin Creek. The old lake bottom is a fine meadow—with little islet mesas of fine shale and sandstone not unfrequently capped with trachyte and rising to about fifty or sixty feet in height clotted over it. The summits being crowned with fir trees was a very picturesque appearance to the meadow. The walls or slopes of the primeval lake are formed of red granite traversed by outbursts of trachyte.

A half hour's ride brought us to Mr. Hill's house and to the mesa whence so many insects had been taken. The mesa stood about sixty feet high: the layers of fine shales nearly horizontal; the top capped with a coarse conglomerate of granite and trachytic pebbles.

We set to work under Prof S direction cutting a section down by means of a trench down the slope of the hill to examine the relative changes of the strata. These were mostly minor in character. The top of the mesa presents a rather puzzling phenomenon. The coarse sandstone being very irregularly bedded and interstratified with pockets of laminated shale. The whole stratification pitches and tumbles in every direction. The nucleus of the mesa seemed to be granite around which as an island in the lake the sediment had accumulated. Around the base of the mesa standing upright as they grew are a number of the famed petrified stumps of a light grey color much resembling the original wood. They average from twelve to fifteen feet in diameter and are said to have stood at one time twenty to thirty feet erect above the ground, till vandals from Chicago came and carted some away and other vandals of visitors cut them down to their present level. They were probably gigantic firs of the *"abies"* class.[120] After our morning work we returned to a substantial dinner at Mr. Hill's and spent the remainder of the day in digging out the fossil insects from the shale using a blunt table knife to open the shales like the leaves of a book or these stone leaves as fine and as white as paper. These wonderful impressions are, as it were, photographed. Amongst it other trophies a lantern fly[121] was discovered. The majority of the impressions being those of ants, beetles, and common flies.

Returning home well pleased with our day after another hearty supper on elk we retired to the bar for a smoke and chat with the Judge. He entertained us with many stories of the explorer Fremont and the men who accompanied him many of whom were French Canadians and fellow citizens of the Judge at his native village of Florissant.[122]

August 20 Was spent by S and myself in a general exploration of the geology of the western extension of the lake. We left Bowditch deeply interested in examining cowpats for beetles especially *Buprestis*[123] and followed the Colorado Springs road in the direction of South Park. We halted first at the base of a bluff by the side of the road of which Dr. Peale gives a report in Dr. Hayden's work.[124] The composition of this bluff was coarser than those at Hill's ranch and the bed of lignite contained many badly preserved specimens of branches and rootlets. Fragments of lava lying on the top were traceable to an overflow about two hundred yards NE.

Thence we followed the branch of the road NW into an open meadow underlaid by shales bounded by walls of granite on one side and a long tongue of trachyte on the other with granite also by the side of it and probably underlying it. We followed the sediments some way up this meadow and finally with the rise of the land they seemed to die out. We retraced our steps by turning up a branch of Twin Creek[125] a little to the South. This branch seemed devoid of sediment. This was explained by what we found further on. That was a great flow of trachyte which must have dammed up the lake till it broke through this obstruction and emptied itself through this channel. Just at this point Prof S who had let his horse graze whilst he went to examine something came back and found his horse gone. Three hours were spent in a weary hunt all over the country. During the hunt I found a fine spear head of agatized wood. Hot and tired we were just giving up when I saw the horse quietly grazing behind a rock a few feet from where it had been left.

After dinner we went over the same ground and made a rude map of the region. This portion of the lake appears to have been divided up into several smaller lakes by floods of lava traversing it and forming dams at intervals of about a mile.

The following day we explored and mapped out the upper or southern portion. This part rolls more than the other. On the East side is a boundary of high granite with alternate overflows of lava. There are also long lacustrine banks running N and S. A long tributary lake runs in on the E with trachyte. Following the lake we come to many mesas and towards its southern end a good deal of cinders and rough scoriaceous [rough fragments of burnt, crustlike lava] trachyte seemed traceable to a conical hill about five hundred feet above the level of the park which was much covered with rather recent looking lava. This may have been one of the seats of the various outflows. Towards the South we could see the outlines of the great Arkansas cañon.[126]

## From Florissant to Colorado Springs; Ute Pass by Moonlight

Leaving our hospitable refuge at Florissant we resumed our journey towards Colorado Springs halting at Haydens Park to lunch and thence passing Bergen Park towards evening we began to enter the Ute Pass.[127] Here in the Fountain Creek Prof Scudder and B proposed a bathe in a beaver dam. Thence we passed down through the celebrated pass the sun had long set but was followed by a bright moonlight which brought out in the grandest manner the outlines of the abrupt cliffs forming the walls of the pass as well as the foaming torrent below. The pass is formed of Lower Silurian[128] limestone and sandstones resting unconformably [a geological phenomenon in which younger strata lie upon older ones with a number of intervening strata missing] upon the underlying granites and rising tier upon tier strata upon strata cut into magnificent towers and castles of red potsdam sandstone and magnesium limestone by the erosive forces. We at last reached Manitou [Springs][129] had supper at a restaurant and took up our abode at the Cliff House Hotel[130] feeling rather ashamed of our rough and disheveled appearance.

## Ascent of Pikes Peak

August 24 Left Cliff House at 7[A]M and took the trail up Buchanans Cañon towards Pikes Peak. We halted on top of the first hill to look down on the magnificent display of upturned strata in the valley below representing nearly every period from the Silurian to the top of the Tertiary rising in waves of strata wave behind wave and with every variety of color red, dark red, variegated yellow, white and brown. Passed over several low hills and reached a mountain meadow or alp with a hut in it which had been abundantly used as a camping place by visitors. The road was very boggy in places. We reached timberline about 3PM and had dinner and then resumed our climb amongst the usual stunted pine and bleached stumps which mark the limit of growth on these mountains. We met three travellers who had walked up but were pretty well played out. By the side of the road we saw a dead yellow bellied marmot or woodchuck.[131] In the crevices of the rock we found a beautiful magenta colored Saxifrage also a bright blue dwarf. The ascent so far had been comparatively easy for our horses but the last three hundred feet to the top was steep and covered with massive blocks of fallen red porphyritic granite boulders. It was in vain to try and drag our horses over there. So we left them at the bottom and climbed to the top. Visited the signal office which was occupied by a sickly looking sleepy young man and two good strong fellows reminding one of pilots with every appliance for keeping themselves warm in winter. As we could get no accommodation here nor anything to eat either we had to give up all idea of stopping on the summit overnight and after a very brief survey of the magnificent view.[132] Enough to notice the valley of the Arkansas and the grand cañon to the South and also to observe that Pikes Peak is cloven on the East by two frightful gorges between 2 and 3000 feet deep very similar to the stupendous cliff that cleaves Longs Peak from summit to timberline. Tired and hungry we hastily dragged our horses down hill fearing to be caught by darkness in the thick timber and in the bright moonlight picked our way home by 9 o'clock.

## Prof Sterry Hunt; Trip to Manitou with Him; Colorado Springs to Denver

August 25 Called on Dr. Solly.[133] Visited Garden of Gods and Glen Eyrie. Dined in Queens Cañon. Rode to Colorado Springs. Stopped at Crawford House hotel.[134]

August 26 Attended church. Mr. Walker preached on *Genesis* and science and denounced Huxley in his sermon.[135]

Met Prof Sterry Hunt[136] in town. A fat good natured jolly but bumptious little man full of talk and very fussy. I took him to Glen Eyrie and Garden of the Gods. He was especially interested and delighted by seeing the juncture of the Silurian with the granitic rocks.

He pronounced the Archaean as Laurentian[137] and carried away several specimens of the red granite some of them without mica. Thence we went to Manitou and the Ute Pass as far as the falls and home to dinner at the Beebe House. Spent the afternoon at Rocky Mountain Club room and breakfasted at Mr. Risley's.[138]

August 28 Bid Sterry Hunt good bye. He went to attend the Association at Nashville.[139] I started for Denver.

Took some notes on road for Colorado Springs.

Took the old road. Examined banks of No. 3 Cretaceous [Niobrara] yellow shales passing this we entered valley with high castellated white rocks which proved to be Tertiary No. 1 white sandstones and concretions of iron. Immediately above it is the Tertiary No. 2 or bed of ferruginous sand and dark ferruginous concretions and ferruginous conglomerate and there at a short distance on the next bluff begin the horizontal beds of the Monument Creek Group.[140] On approaching this group you first meet debris of hard ferruginous conglomerate which are usually found capping the monuments and are probably here found left as the only hard parts of monuments that have dissolved away. The composition of the monument resembles that of the overlying Tertiary and is derived from the same source viz the granitic Archaeon rocks. Whilst we were studying these rocks two magnificent eagles hovered over as in easy range. These majestic birds will often approach very near and seem either destitute of fear or else despise it. We were soon among the monuments of every form and shape, the flat topped mushroom being the most common, the base broad and gradually tapering upwards towards the neck which is capped by a quartzitic conglomerate cemented together by iron. Passing through the labyrinth of these, I rode on up the valley with groups of these in various conditions, some forming solid castles with incipient monuments branching off from them their tops not unfrequently still united by the iron rust.

We halted for the night at a little hotel called Monument House where the valley heads up towards what is called the Divide.[141] The people of the hotel told me that a certain very learned geologist had boarded there and showed me a fossil carefully wrapped up with an extensive label stating that it was a bone which had been found beneath so many thousands of feet of rock and was many millions of years old. The two latter points I did not contest but the former I did for this interesting bone proved to be nothing more than a piece of petrified palm wood.

The following morning I climbed one of the magnificent castles of the Monument Creek Group which commanded the top of the divide, a fortress of yellow sandstone flashing out among the fir trees. It stood on a hill about 150 feet. Around the base of the hill lay many fragments and loose boulders of the same rock though at first glance from their being granite fragments worked over again into a course granitic conglomerate. They resembled huge masses of original granite of a very coarse kind.

The castle is composed of coarse sandstone passing into a yellowish conglomerate with argillaceous [resembling or made of clay] concretions and bands irregularly dispersed through the mass. The pebbles of the conglomerate represent every granitic type quartz both angular and round with flesh colored feldspar predominating. The top of this monument is rounded into many nipple like forms and the mass is divided up perpendicularly into and horizontally into forms resembling chessmen; thus

The horizontal dividing lines are usually at the softer points where one stratum lies on the next.

To the right, west of this monument is a park displaying some of the red and white Triassic monuments. A series of terraces divided by creeks and ravines and covered with timber come down from the mountains and pause at the open plain between them and the "panniers" to the East which form the continuation of the Monument Creek Group a distance of about two miles or more, forming probably the basin like drainage of some heavy body of water in the later periods of primeval days.

The Dakotah group and Cretaceous beds seem covered for some miles along the base of the mountains.

Standing on this natural fortress we seem surrounded with fortresses high yellow sandstone ramparts fortifying the North slope of the top of the Divide and commanding the valley with its towers and portholes, it seemed as if we could imagine cannon to right of us cannon to left us suddenly blazing out on the enemy.[142]

The mountains descending behind the Red Trias are of a deep salmon colored red from the red feldsparthic granite. Bergen Park lies behind these foothills.

The elevation of the Divide we should estimate roughly at about 6000 feet and the highest monument or castle not over 200 feet.

Opposite our fortress was another singular structure resembling a natural portico of a Greek theatre. This portion of the Divide above Monument House and station seems the line of separation between the ancient erosive forces dividing in later primeval times the waters that are on this side from the waters on the other side by a wall of sandstone 300 to 400 feet thick. The lower portions of which are now covered with grass.

The top of the Divide is composed of a series of mesas or table lands separated from one another and from their summits. You look down on the Northern slope into an extensive basin with high mesas rising from its bottom in every direction and marking or registering the high line and limit of the ancient plain thus:

To the East this group indicated by the pine forests covering it stretches away to the horizon, split up here and there into many soft basins of smooth turf with here and there a stray fortress or mesa.

Looking North, is still more picturesquely broken up into fortress-like table lands, mesa rising above mesa are lost in a labyrinth of distant mesas whilst pretty little parks of green grass sparkle out in the sunshine from the dark scrub oaks with which the lowlands and slopes of the table lands are thickly covered and which in autumn must present a gorgeous appearance when this foliage is turned into scarlet and gold. The trunks of these trees rarely attain a greater thickness than ten inches diameter. To the North and South stretch the grand ranges their sharp snow capped peaks strongly contrasted with the flat mesas at their base. Pikes Peak commands the view of the range to the north and Longs Peak to the South like two guardian sentinels with the rather even topped outlines of and undulations of the foothill range enclosing Bergen Park on the West. The mountain forests were on fire in many places and their smoke obscured the landscape to the North.

We descended into the Basin and climbed to the top of one of these singular mesas called Lookout Mountain near a station called Greenland.[143] The mesa is about 700 feet high capped with fine, pink, reddish, brown, and gray trachyte massive and splitting with a conchoidal fracture [revealing rocks with shell-like surfaces]. The thickness of this is from twenty to thirty feet. The top is broken up into flat slabs which ring under the hammer. The lava is slightly vesicular in places. Some of the sandstones of the underlying conglomerate are enclosed in the lava and in some places fragments resembling fossil wood are enclosed. This massive lava top is underlaid by a very coarse conglomerate of granitic pebbles of a somewhat harder and coarser texture than those composing the monuments on the other basin and protected by its cap of trachyte. These mesas resisted the erosive forces which debased the beds on the North slope of the Divide. In the banks of a stream below Lookout Mountain seams of dark lignitic clay with impressions of reeds or grass overlying coarse conglomerate were observed. Lookout Mountain was much used by the Indians to view the landscape over and their arrowheads are found on the top.

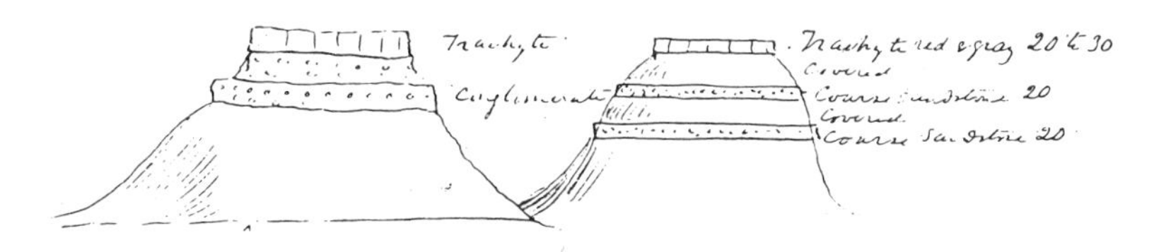


From the top of Lookout Mountain we look down on a smooth basin like meadow occupied by several neat little farms, their paths, fields, and crops marked out like a map of the United States and its railroads.

This Divide country is noted for grazing cattle. Many herds of these were browsing below.

Following the road to Denver along Plum Creek we pass between and under a perfect labyrinth of these mesas all remarkably similar in shape and structure to one another. It seemed strange to be traversing the by ways and paths several hundred feet below the original surface outlined by the tops of the mesas.

The weather was intensely hot and we grew weary of the monotonous table lands and scrub oaks and were glad towards evening to catch sight of the box like top of Castle Rock rising in the distance, our destination for the night. The land this side of the Divide is much more broken up and rolls more than that on the other.

## Birds and Squirrels; Castle Rock; Hotel Politeness

A good many large marsh buzzards[144] swooped over the plain and Beau was often attracted by the shrill bird like whistle of the tiny spotted gopher *Spermophilus tridecentineathus* running between the artemesia bushes [sagebrush], on standing erect like a prairie dog to get a better sight above the grass of coming danger. It is worth noticing here how many of the prairie animals are endowed with this faculty of standing erect. Otherwise from the low creeping habits of these creatures they would have no means of seeing approaching danger on the low grass covered plains.[145] The streams this side of the divide all flow into the Platte in the other into the Arkansas.

I reached the little village of Castle Rock about sunset tired, dusty, and hot. Rode through the single street to a little white country inn. The hostess, an Englishwoman, by the way taking in my rough appearance and thinking the Rev. Professor was probably one of "the Black Hills fellers,"[146] my geological hammer in my belt looking suggestive of revolvers and so forth, informed me that they were quite full and could not accommodate me at her Otel, a pardonable hotel lie; but that I could find a H' Inn across the way. I retired to a humbler manor where they treated me well in a groat house [i.e., cheap lodgings] with mortar walls but clean and the good woman after a supper of eggs and bacon kindly sewed up for me certain rents in my garments an 'owdashus' liberty I could never have asked of myne hostess of the Otel.[147] Next morning I ascended the Castle Rock close to the town. It is a box shaped mass of conglomerate sandstone on top of a hill about 600 feet high. The box top is composed of very coarse conglomerate of very large pebbles banded by seams of finer conglomerate sandstone irregularly bedded. The stones forming the conglomerate are granitic pebbles, the quartzose element predominating with very large grey trachytic pebbles profusely interstratified with the mass at all heights. Of it some of these pebbles or masses are a yard square and are of the same gray trachyte which forms the upper flows and cap the mesas usually. . . . Castle Rock has no trachytic top. The presence of these trachytic masses proves that volcanic flows had occurred previous to the deposition of the sandstone and conglomerate and previous also to the final flow which caps the mesas and that the conglomerate was deposited in rough shallow water gathering these pebbles into the sandy mud. The conglomerate at the rock was coarser than anywhere else. I noticed on the Divide further down the hill is another bed of finer conglomerate.

View from Castle Rock: South the basin of 'mesas' shut in on west by foothills above which rises with craggy outlines, Platte Mountain;[148] Pikes Peak, southwest solitary gray and misty with the foothills forming a wall. Bright little village of Castle Rock with its new yellow houses and white cottages flashing on the level below in the morning sunshine. This is the county seat and numbers from thirty to forty houses, stores, two hotels, a state [building] and school house, and church.[149] Plum Creek forks near here one branch turning off to the SE. Looking North we have the other side of the basin shut in by a crescent of mesas. To the W and NW the main range which seemed lost below Pikes Peak is resumed by Mt. Evans[150] and after that the range rises into a chopping sea of snowy peaks.

Just where the Range is resumed South Platte cañon emerges on to the plains passing through a great development of the red rocks of the Trial.

It is possible that much of the pebbles found scattered over this portion of the plains are derived from the breaking up by denudation of the conglomerates of the Monument Creek Group.

The tops of the mesas indicate an erosion of from 600 to 1000 feet of thickness of material over the entire region. I rode for some hours along the edge of these mesas till gradually they dwindled down and passed off into the level plains as we neared Littleton[151] and the Platte river.

## Crossing the Platte; Farming; Back to Morrison

At a point facing Platte Cañon I struck across country toward the Cañon with a view to a closer examination of the Cretaceous hogback of the Dakotah group.

Soon after crossing Plum Creek we found ourselves in the bed of the Platte. Here I halted to bathe and dine and then floundered on through marshes to try and find a crossing over the river. Jenny as usual declined to cross anywhere. At length we forded at a place where bars led into a field in which a large party of farmers were thrashing grain with a steam thrasher.

> "Huzzin and maazin the blessed fields
> with the Divils own team"[152]

Crossing the Platte, we passed through the rampart of the Dakotah sandstone and entered the foothills by Deer Creek. Here was a good exposure of Cretaceous and signs of bones.[153] We rode between the ravines now and then starting a rabbit to Beau's delight till we reached Morrison late in the evening where we found abundant letters awaiting us and learnt that Prof Brackett of Princeton[154] had called at camp in our absence and that Prof Mudge was still at Canyon City. Rode up to camp and found George and Nugent[155] awaiting me.

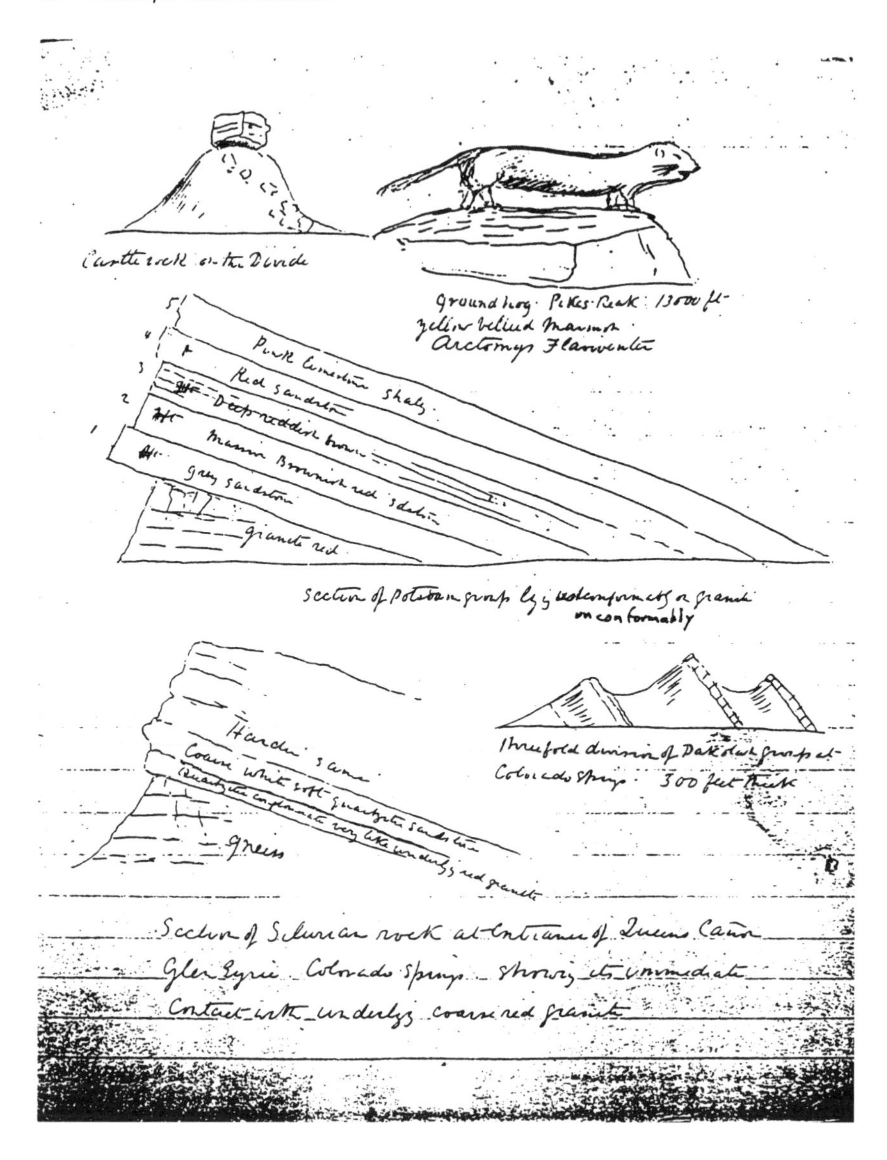


In Glen Eyrie, Queens Cañon is cut to a depth of 700 feet through massive red granite and rocks of the Lower Silurian. Upon the massive red coarse granite which is sometimes only red feldspar and quartz without mica or hornblend about 600 feet above the stream, the Silurian rocks lie uncomformably. The layers lie evenly one upon the other, the lowest weathering to a light gray with rugged outline above this is a dark reddish brown massive sandstone. The layer above is deeper red but less massive. The top is capped with gray shales. There are three miles of rock thickness from Colorado City to Manitou. From the Niobrara limestone to the Silurian and allowing for dip about 12 to 15,000 feet thickness of strata.

The divide was evidently once the top of high plain. We ride along on what we call the plain, but it is really in the deep valleys which have been washed down to a hundred feet below the original plain. The mesas are illustrations that it is water that has made mountains by washing out valleys. The lava capping these must have poured out over them from some source to the north probably through cracks in the earth. [See facsimile, p. 62.]

## Morrison and Bones Again

Next morning I went with the boys to see what they had found in my month's absence. They had got out a quantity of rock and amongst it some very fine specimens, huge black shafts of leg bones reminding one more of stove pipes than anything else were lying in bas relief exposed on big blocks of sandstone weighing a ton or more whilst the butt ends of other bones were protruding from the stone; teeth and spines and plates of armor were shining on other fragments and fragments of various kinds of bones lying around in various conditions promising me a long and tedious work in whittling down the big blocks and in matching together the scattered fragments.[156]

September 1 Spent in whittling out a large limb bone from its stoney surroundings and in doing so it broke up badly like biscuit. Nugent meanwhile put in a blast and exposed a lot more. George I employed in box making at the bottom of the cliff. In the afternoon it rained heavily and when it cleared up we built a covering for our spring under the tree.

September 2 It rained all night and all day and we spent a rather comfortless day huddled together in our tent and cooked our rice pudding and raisins under some difficulties and eat it on a gunny sack in our tent with our legs curled up under us like cobblers.

## Large Spines or Ben Butler's Spoons

Our work continued after this day by day without any exciting novelty beyond the discovery of some new portion of the skeleton. The most remarkable was one evening when I came home from a trip to Idaho

Mr. Shields called me to come and look at what he insisted were spoons.[157] They were three or four long cylindrical and ck fragments resembling the ends of spearheads or long spikes with a spoon like termination at the thicker or butt end. These I conjectured were some of the long dorsal spines, smaller ones of which we had before met with. Some of the blocks of stone I found it necessary to pack entire as any attempt to break them open and get out the bone would have resulted in fracturing the bone all to pieces beyond any hope of putting it together again. To transfer these heavy masses down hill we had to construct a sort of sleigh and the whole party would carefully draw the big fragment like a man with a broken limb down hill near the road where a box was made for it, and it was packed with care in gunny sacks and hay. Soon we had a row of big blocks and big boxes ranged at the bottom of the hill and we sent for a wagon to convey them to town and to the railway station. In this way I used to send off over ten boxes a week averaging 250 pounds a piece and many of them from 4 to 600 pounds. With September the intolerable heat of August and July began to abate and work became much pleasanter. Soon the flowers began to fade in that graduation of characteristic species so noticeable in these mountains where a species seems suddenly to burst forth and cover the ground with white or blue in a few days such as the *Leucocrinum montanum* [sand lily] or the *Oenothera* [evening primrose] or the *Pentstemon* [beardtongue] and then pass off to be succeeded by some other species quite rapidly, many of the old varieties still lingering here and there and some continuing till the end of the season reminding one of the manner in which a few species of one geological epoch run into those of the next whilst the main types are passing away to be succeeded by higher forms. A few relics still remain and live on with their new creation to tell of days gone by and to show that abhorrence of sudden change which is characteristic of Him who changes not.

## The Fall Season and Autumnal Foliage

The bold glaring faces of the sunflower faced about towards the sun till the very last and timed as to our work at sunrise and sunset. Soon the flowering shrubs faded and their places were supplied on the rises by a few small gooseberries and some large yellow and black currants of a very course flavor. The fruit of the choke cherry also hung in purple clusters from the bushes. Then a still sweeter period stole imperceptibly upon us towards the close of the month. The beginning of the Fall was ushered in by a delicious coolness of the air, a slight frost in the morning involving an additional blanket, that solemn stillness in the air broken by the tinkle of a falling leaf, and a slight yellow beginning to creep over the foliage which soon deepened into brighter colors and bright patches of gold and orange flashed out from the little hollows and ravines of the mountain above as with whom the season had begun earlier than in the few hundred feet below. Soon the most gorgeous red and scarlet began to tinge the edges of the mountain sumach *Rhus* first encroaching on the green then dyeing the whole fernlike leaf a deep blood color and from that to a bright scarlet. The poison ivy also vied with it in brilliant orange and gold and various little bushes of the dogwood, maple, plum, and other trees assumed every variety of hue, all however maintaining that soft autumnal gradation so in unison with the refinement of the season. It was not long before our camp was invaded by ladies on horseback from the hotel at Morrison who came up to gather the leaves to paste up in graceful festoons around their rooms in Denver. Sometimes these ladies would stop and partake of a cup of coffee or a glass of lemonade at our tent. And my office of a morning was usually to explore the neighboring ravine whilst the boys were preparing breakfast in search of leaves to send to Denver to friends or to give to our guests at the Evergreen Hotel. About this time the election took place and Mr. P[158] of the Hotel being a candidate for the position of Justice of the Peace. My men were frequently canvassed, on one occasion by a young lady of the hotel who walked out all the way for the purpose: but my men were more strong minded than Adam although they did think sending a lady to canvass was coming it pretty strong and scarcely fair play.

## Sunday and Animals around Camp

Sunday morning whenever not called away by duty[159] was very enjoyable. We all lay in bed much longer than usual and the various birds and animals of the vicinity took advantage of this to explore our camp more accurately than on other occasions when we were generally up and about as soon as they were. Soon after sunrise I was generally awakened by the loud harsh cries of a couple of magpies *(Pica melanolence)*[160] perched on the ridge pole of my tent and inquisitively looking down at me through a small rent in the canvas. Not unfrequently one of them would lose his balance and slide head over heels down the smooth canvas with cries which I am convinced in bird language were as near profane imprecations as it is possible for birds to utter. Then a large spotted woodpecker with a red crest[161] would swing himself by an easy jerky flight from tree to tree and utter a screech of exultation as he found camp forsaken and a good prospect of a savoy meal where greasy bacon had been left hanging against a tree. He would poise himself against the bark using the short feathers of his tail as a fulcrum and swinging his head back and forward with the powerful pulley like muscles which run from the slender neck over the back of the head deliver such tremendous blows with his beak that one wondered both beak and skull were not fractured at each crack. But I noticed he never gave a full blow right in front of him but always worked around the back by a series of sidelong taps verging towards the centre of the hole which he was rapidly excavating. One morning whilst we were getting at breakfast he beat a regular devil's tattoo for us on the frying pan so close to us we could have touched him, but quietness and Sunday morning breeds wonderful tameness even in Rocky Mountain Birds. Another little bird the chewink [one of the towhee species, *Pipilo*] used to creep and rustle and scour up the dead leaves under the bushes so much like a rat or squirrel that we often mistook him for one.

About the end of September I received a telegram from Prof Marsh desiring me to go down to the excavation nearer Morrison and see if I could find any more of our monster [the *"Atlantosaurus immanis," Apatosaurus ajax*] that we had left there and whose excavation we had abandoned after the accident of the falling rock. Accordingly we intended to move our camp to a little broken down shanty nearer the work.[162] A night or two before making this change a little adventure occurred. I was sleeping alone in my tent when there was a rustling in some paper close under my head. I jumped up and ran out of the tent when to my disgust who should I see emerging from the bottom of the tent but a large skunk and presently he ambled off under the moonlight down over the bank into the stream bed, fortunately I had presence of mind not to disturb or pursue him. The next day, Beau our shepherd dog, was left tied to the pole of the tent, an insult which he resented by tearing the front of the tent all to pieces and we found it on our return lying in shreds on the ground. So that night I had to sleep in the open. The night was very lovely and a bright moon. I was awoke about midnight by some animal's feet pattering close by my head I jumped up supposing it was my dog got loose when there in the moonlight was my friend the skunk ambling along gaily a few feet from my bed. I was mad and disgusted at being awoken up at such an hour and by such a creature and without thinking of what would be the result, I cried out "Take that you brute!" and slung the first thing that happened to be near me, which was an axe, at the intruder. At once he paused as is their habit, elevated solemnly his fine bushy tail to a perpendicular and then gave me a volley of fine mist which though he was full twenty feet from me fell all over my bed and blankets as I soon had cause to know from the intolerable stench. Now that I had been and gone and done it, I determined to finish the matters by a volley of stones which soon knocked my gentleman over the bank and down into the ditch where I piled stones on him until I was quite sure of his death. But who shall tell the nasal treat through which I passed till morning. Moreover it was rather trying when next day I went down to call on a lady in town and as she was talking to me, she observed sniffingly, why dear me there must be one of them things around again. I dared not confess who them things were. I changed my clothes and blankets, put some earth on them, and left them for a long time to air ere I again enclosed them.

## Go to House Keeping; Hammock

Our new cabin needed a good deal of repairs. First we had to make a board bedstead for the men to sleep on and various articles of furniture such as a table [and] bench to sit on. Moreover in endeavoring to repair the gable end of the building the whole gable fell in and had to be replaced by a partition of lumber planks covered with the remains of our late tent. Then the roof had to be patched and a stove and stove pipe erected, shelves for our library and other knickknacks. After a time things began to look quite homelike and cheerful as we sat down to our first breakfast under a solid roof house. George, whose sorrows as a cook over an open fire, highly approving of the change. For myself I still adhered to my tent life and adjourned every evening to my canvas dog tent pitched by the side of the stream. At last however after one or two snow storms it became rather dismal to leave the warm stove and go out through the snow to the cold tent so I came in and for the first night slept on the floor but the mountain rats kept up such a racket under my head and about the room that sleep was impossible. So I turned my tent into a swinging hammock and swung myself on to the rafters above the boys head. At first it was a ticklish matter climbing in. I had to put my foot carefully for fear of treading on the noses of the sleepers below, then balancing one leg on a bit of scantling, and swinging the other in the air like a trapeze dancer, I would swing myself into the hammock, those below in considerable anxiety as to my success for fear of my coming down on top of them hammock and all with obliterating effect. The tremendous bulge in this ex tempore bed formed by my body excited a good deal of satirical merriment from those below. But there is no bed so comfortable as a hammock when once you have got in to it.

By and by our leaves and our ladies faded away together. The weather grew colder and our first snow storm occurred quite early at the end of October. Meanwhile our excavation went along with great success. We soon blasted and removed the fallen rock and underneath it discovered abundant signs of bones. In one place we came upon a number of vertebrae and amongst them a colossal mass which proved to be the sacrum consisting of three vertebrae coossified and weighing over a hundred lbs. in weight close to this was a curious large scimitar shaped bone (a scapula) two or three limb bones.[163]

But the most gigantic I may say prodigious bone was discovered by one of my men. He was removing a portion of the bank when he came upon a mass which from its size and roundness might have been taken anywhere for a big loose boulder. The more he removed the earth the more huge grew its proportions. At last he uncovered the whole of it and rolled [it] out on to the bank, a portion of a big butt end twenty-five inches for diameter, a little further search revealed other portions of the shaft broken into sections and the butt and five feet of the shaft of a gigantic femur were section by section put together and exposed on the platform of the bank like the broken column of a old temple. Nothing could give any one a greater idea of the vast proportions of the animal than this fragment of which at least four feet forming the other end were missing and could no where be found.[164] On the following Sunday there was a general rush of visitors to see this prodigy and those who could get over their incredulity as to its being a bone at all, an incredulity which was most pardonable, tried the experiment of how many persons could sit comfortably on it at once. Our little party of three used it daily as a bench to sit on whilst we eat our frugal dinner of beefsteak and apples cooked on the spot.

About this time our party was reinforced by the arrival of Mr. Williston[165] one of Prof Marsh's assistants and a very pleasant fellow who had been the companion of Prof Mudge in his trips in Kansas. We entertained him as well as we could in our little cabin and our evenings were made pleasant by stories and by an abundant supply of *Harper's Weeklies* sent to us by the thoughtfulness of Prof Marsh. We were all at work together one day under the overhanging ledge and at evening laid our tools as usual in the hole and quit work. During the previous day it had snowed heavily, and this day being warm, a rapid thaw followed. About midnight one of our men was walking through the village when he heard a roar like thunder from the cliffs above as if the whole hogback was coming down. Next morning we came as usual to our hole when lo there lay the whole ledge fallen in and filling the excavation with a weight of rocks of over 100 tons many of the blocks weighing singly upwards of a couple of tons. Had this fallen when we were there our entire party would have been crushed atoms and buried beneath tons of rocks which afterwards took us over week to remove by blasting and sledge hammers. Several of our specimens were crushed and our tools also. This was accident number two and made us more careful for the future. During November the storms of snow were more plentiful but they did not deter us from our work as we could dig with comparative comfort under the shelter of the ledge, for by this time we had run our excavation full forty feet into the side of the hill. The snow was an advantage to us in another way as it enabled us with greater ease to draw our heavier specimens in a rude sleigh down hill to the boxes. We had however some pretty cold snaps. The thermometer one morning sunk as low as twenty below zero and we had to keep a fire burning in the hole for the men to warm their fingers at. We not unfrequently cooked and ate our dinner on the snow. In the evening we sometimes toiled home through slush and heavy mud or heavy snow to our cabin and were welcomed by a chorus of coyotes in full cry issuing from the mountains on a night's foray. This made the sense of wildness and solitude all the more striking.

## Remove Camp to Morrison

We generally knew when there was going to be a snow storm by the hooting of the owls in the red rocks the evening before. Also not unfrequently these birds would give their hoo hoo notes early in the morning. At last our little cabin grew so comfortless and the road to it so long and tedious that we concluded to go down into Morrison and take a little cabin there close to our work. Here we passed the winter months of December and January and by end of January Prof Marsh brought our excavations to a close as there was little prospect of any more bones. He together with ourselves were much disappointed in not finding any skull which would have enabled us more closely to identify the Dinosaur.[166] We found it a much more sociable life in Morrison than our very cheerless and lonely one at the nest. We experienced that it is not good for man to be alone. We passed many pleasant evenings at the hotel,[167] and attended some amusing theatrical entertainments at the school.[168] I gave a couple of lectures there on the Dinosaurs to the people who had often made many inquiries about the bones and the animals to which they belonged and had generally taken much interest in our work.

I tried to sum up all the questions that I could remember to have been asked me and answered them to the best of my ability such as: What kind of an animal was he, how long, how did he get into the rock and how did his bones become turned into stone. One old gentleman, a colonel who came out on the Omaha excursion[169] was brought puffing up the hill to see the wonder and asked me out of breath, "Mr. Lakes I don't see how an animal seventy feet long ever climbed up this hill when I can't do it myself?"[170] I frequently had to arm whole parties of people up to the diggings and standing on the edge of the dump explain to them the history of the strata of whose development wave upon wave there was the grandest example before us. It was easy from this shot to point to the prairie and show how the strata had once been laid out flat at the seas bottom and how as we approach the mountains the uplifted edges show the upheaval of the mountains, carrying the stratified rocks up with them. Then the age at which this animal lived was asked and I could reply by pointing to the thickness of the rock above them over 300 feet every layer of which might indicate many hundreds of years and then to the vast thickness represented on the prairie of a mile or more of rock all of which lies above our Dinosaurs. Then I could also give an idea of the dimensions of the creature by pointing to the length of the excavation which was about 50 feet and to a rock above us which gave his altitude when standing erect at 35 feet.

## Description of Fossiliferous Localities and Life in Kansas (from Cope)[171]

The country of Kansas in which so many of the mosasaurid reptiles, birds with teeth and plesiosaurs have been taken is Niobrara Cretaceous between Missouri and Rocky Mountains and is of a rolling prairie character. It represents the old bed of seas and lakes so gradually elevated as to suffer little disturbance unlike our tilted hogbacks near the Rocky Mountains. These sea and lake beds have not been pressed into hard sediment by superencumbent rock nor roasted and crystalized by internal heat or metamorphism. The rock is limestone yielding to action of water which has cut them up into labyrinths and cañons. The stratum is a yellow chalk and is cut up into all the bad land scenery. Huge oyster shells some opened and others with both valves together lie scattered over the ground like the relics of a meal of some titanic race twenty six inches across. *Haploscapla*.[172]

At the bottom of the gulches in the rain washes, the explorer picks up a tooth or jaw and will trace these to a bank or bluff where lies the skeleton of some ancient sea monster. The vertebral column runs into the sandstone with a paddle extended on the slope. Sometimes a pile of remains will be discovered which the breaking up of the rock has deposited on the lower level. How came these here and what kind of creatures were they when alive. They lived in Cretaceous epoch when the chalk of England and greensand of New Jersey were deposited and many reptiles and fishes peopled the sea covering the land at that period. Thirty-seven species of reptiles found in Kansas varied from ten to eighty feet and were of six orders only one was a land animal four were fliers the others inhabitants of the sea. They swam over the plains from Arkansas to Fort Riley [Kansas] traversed Minnesota to British possessions and Lake Superior. The extent of the sea to the westward was vast. It was probably a shore now submerged beneath the North Pacific. For description of some of the animals see printed account.[173]

The[y] were wonderfully long, heads large and flat, conic eyes directed upward. Two pairs of paddles like flippers of a whale with these they swam with speed. Four rows of teeth mainly for seizing their prey more than for cutting or mastication for they swallowed their prey entire like snakes instead of that wonderful expanse of throat in snakes which is due to an arrangement of levers supporting the lower jaw. Each half of the jaw was articulated midway between ear and chin. This was of the ball and socket type enabling the jaw to make an angle outward and widen by much the space between it and its fellow and here we may see how a scientist goes to work in reconstructing animals from slight materials.[174]

The habit of swallowing large bodies between the branches of the under jaw necessitates the prolongation forward of the mouth of the gullet. Hence the *Pythnomorph*'s [Cope's name for mososaurs] throat must have been loose as a pelican's. Next the same habit must have compelled the forward position of the glottis which is in front of the gullet. Hence these creatures could have uttered no sound but a hiss as do animals of the present day which have a similar structure as the snakes. Third the tongue must have been forked and long its position still behind the glottis so no space for it except enclosed in a sheath beneath the windpipe when at rest or thrown out beyond the jaws when in motion.

The *Pythnomorphs* of Kansas were *Liodon proriger, Liodon dyspelor*. Seventy-five feet it had a long projecting muzzle like the blunt nosed sturgeon. It may have used this snout as a ram. *Liodon dyspelor* was the longest of known reptiles.

Account of the Discovery of *Liodon*[175] A part of face with teeth was observed projecting and we attacked it with picks and knives. The lower jaws were uncovered with glistening teeth also vertebrae and ribs. The delight of the party was at its height when the bones of the pelvis and hind limb were laid bare for they had never been seen before in this species. Whilst lying at the bottom of the Cretaceous the carcass had been dragged hither and thither by sharks and other animals and parts of the skeleton were displaced and gathered into a small area. The massive tail stretched away into the bluff and after much excavation the most of it was obtained.

Of *Platecarpus Coryphaeus*[176] After examining the bluffs for half a day without result a few bone fragments were found in a wash. Others led the way to a ledge forty or fifty feet from both summit to base where stretched along in the yellow chalk lay the projecting bones of the entire monster. Several vertebrae were found protected by the roots of a small bush and when these were secured the pick and knife were used to remove the remainder. Just then one of the gales so common in the region sprang up and striking the bluff reflected itself upward. As the pick struck the rock the limestone dust was carried into nose and eyes and every available opening in the clothing. I was blinded and my aid [eye glasses] disappeared in the cañon. A handkerchief tied over the face and pierced by holes kept me in total blindness. But a fine relic of creative genius was extricated from its ancient bed and one that leads its genius in size and explains it structure.

On another occasion riding along a spur of a yellow chalk bluff we saw some vertebrae lying at its foot. Examination showed that the series entered the rock and on the other side the jaws and muzzle were seen projecting though laid bare for the convenience of the geologist. We speedily removed its blocks to the level of the reptile and took out the remains that lay across the base from side to side.

*Clidastes*[177] Not so large as *Liodons* and elegant and flexible. To prevent contortions from dislocating vertebral columns, there were an additional pair of articulations at each end whilst their muscular strength is attested by elegant striae [grooves marking the connection of muscles] on their bones. Forty feet long A smaller one called *tortor*[178] was of elegant proportions. The head long and lance shaped its lithe movements brought many a fish to its knife shaped teeth it was found coiled up beneath a ledge of rock with its skull lying undisturbed in the center.

Flying Saurians[179]

| | |
|---|---|
| *Petrodactytus occidentalis* (Marsh) | 18 feet |
| *Pterodactytus ambrosius* | 25 feet |

These flapped their leathery wings over the waves often plunging seizing many an unsuspecting fish or viewed at safe distance the sports and combats of the saurians. At nightfall they hung suspended to the cliffs by their clawlike fingers of their wing limbs.

Tortoises[180] Were the boatmen of the Cretaceous but none had been known from Kansas until recently. *Protostega gigias:* the house of this creature is formed by the expansion of the usual bones of the skeleton till they meet unite and become continuous. The lower shell is formed of united ribs of the breast and breast bone with bone deposited in the skin. In the very young tortoise the ribs are separated as in other animals. As they grow older they begin to expand at the upper side of the upper end and with increased age the expansion extends throughout the length.

The fragments of *Protostega* were seen by men projecting from a ledge of a low bluff their thinness and distance to which they were traced excited my curiosity and I attacked them with the pick. After several square feet of rock had been removed, we cleared up the floor and found ourselves repaid. Many long slender pieces of two inches width lay upon the ledge evidently ribs with the usual heads but behind each head a plate like a flattened bowl of a huge spoon placed crosswise. Beneath these stretched two broad plates two feet wide no thicker than binder board, edges fingered, and surface hard and smooth. All this was new to us, after picking away the bank and carrying the soft rock new masses of strange forms were disclosed: bones of a paddle and a leg bone, the shoulder blade of a huge tortoise, and we knew we had stumbled on the burial place of the largest sea turtle ever known. Single bones of paddle: eight inches giving the spread of the expanded flippers at fifteen feet. The ribs were those of an ordinary turtle just hatched. The great plates represented bony deposition in the skin but it was incredible that the largest known of turtles could be just hatched. It is one of those forms not uncommon in old days, whose incompleteness points to a truth that animals have assumed modern perfections by a process of growth from simple beginnings. The western Cretaceous seas had remarkable fishes, sharks were common, also the salmon and saurians.

Vertebrae and other fragments project from the worn limestone. There was one whose bones crowned knobs of shale left standing amid surrounding destruction. The head was some inches longer than a full grown grizzly bear and the jaws were deeper in proportion to their length. The muzzle shorter and deeper than that of a bull dog; teeth cylindric fangs smooth glistening irregular at certain points in each jaw. They projected three inches above the gums [and] were sunk into deep pits long as the fangs of a tiger but more slender. Two pairs of such fangs crossed each other on each side of the end of the snout: *Portheus molossius*.[181]

The ocean in which these flourished was at last completely enclosed by elevations of sea bottom only communicating with Atlantic and Pacific at the Gulf of Mexico and Arctic Sea. The continued elevation of both Eastern and Western shores contracted its area and when ridges and sea bottom reached the surface forming long low bars parts of the water area were enclosed and connection with salt water prevented. Thus were the living beings imprisoned and subjected to many new risks of life. The stronger could more readily capture the weaker while the fishes would gradually perish through the constant freshing of the water. With the death of any considerable class the balance of food supply would be lost and many large species disappear from the scene. The most omnivorous and enduring would longest resist the approach of starvation but would finally yield to inexorable fate. The last one caught by the shifting bottom among shallow pools from which his exhausted energies could not extricate him.[182]

## On the Science of Paleontology[183]

The first law of this science is the uniformity of nature's methods and the persistence of type. An organized structure once created and meeting with no opposition is adhered to and extended in time and space. On this basis the forms of the past are reconstructed. Persistence of type presupposes a knowledge of pattern. Patterns quite distinct from those more known to zoologists have existed in past ages. How can the structure of a species be inferred from a fragment? Our ignorance of the exact end of a line or series is a difficulty in our way.

If a fragment of an animal be found with a certain type of teeth of a *(Selenodont)*.[184] The law of uniformity of type is that the first bone of the hind foot possessed two fully grooved faces one above and one below, and not only as in animals, and also that the lower pulling face was succeeded by two subequal toes and that the lateral toes were reduced or wanting, from the tooth we judge the hoof. Again.

If I find a part of the structure just with the first row of bones united into one mass and closely embracing the leg bones without being continuously united I know I have an animal with teeth with a very long hip bone and a long series of united vertebrae or sacrum resting upon it; in other words a Dinosaur.

There is also a law of successional relation.

The foot of a dinosaur is intermediate between a reptile and a bird, so are the sacrum and pelvis.

Again long legs in a grazer suppose a long neck to enable it to reach the ground with its lips. Hooked claws suppose carnivorous teeth or a hooked beak.

For a horizontal body to be poised on two legs instead of four the weight of viscera must be transferred backward and anterior regions of [the] body lightened. This is the case with birds and Dinosaurs the lower bones of the pelvis are thrown backward the fore limbs lightened and the head reduced in proportionate size.

## Relations of Paleontology to Geology[185]

Its relations to Geology are empirical.

Its indications are definite for one locality but not identical for all localities on the earth's surface. The lower we descend in the scale of being, the more uniform over great areas are its phenomena.

Among higher animals the geographical peculiarities are greater than the stratigraphical. Hence some think fossil vertebrates cannot give conclusive evidence of age of rock strata, for in Australia there are animals approaching more nearly to those found fossil in the Jura of Europe than to any now living. So that of Australia was submerged and her fossils again brought to light we should say the sun had never shone on it since the Jura.

The earth now represents four distinct faunal areas: Australia, South America and temperate lands of northern hemisphere. Each possessing peculiar forms of life not now found elsewhere. This distinction has always prevailed. The faunal distinctions have a very ancient origin and are to be first considered when estimating age of strata from mammalian remains. An advance has been made by discovery of great similarity between extinct forms of northern hemisphere and living or modern ones of the southern hemisphere faunae. Australian fauna are Jurassic, South America and Africa related to Eocene.

How is life significant of chronological station in earth's strata.

Since very many forms of animals are widely spread and also distinctly limited in range on the earth's surface the same order must have prevailed in past times and have been of equal significance. The same holds in paleontology. The apparition of types over the northern land area has been nearly universal. The succession of Tertiary beds with mutually similar poraneous deposits contained on a large degree similar life. So of taceous.

as this succession of life universal over the globe and do these trenchant lines justify the old assumption of repeated destructions and recreations of animal life. The first question is negatived by the explanation of the character of existing fauna of southern hemisphere. Immigration from one continent to another has taken place sufficient to account for abrupt changes in the life of each without intervention of creative acts. The glacial periods 25,000 years apart caused migration. The life of both northern and southern hemispheres are not homogeneous. Elephant is Pliocene and rhinoceros Miocene. In North America the opossum and raccoon are Eocene. Wolves and foxes are Miocene, weasels Pliocene. Cats appeared in the Pliocene.[186] Man signifies the glacial period and reaches his culmination in the ages that intervene between that great time boundary and one to come.

Thus a proportion only of life of an epoch characterizes it and originates in it, remaining members being legacies from preceding ages. The lastest forms of life embraced in an extinct fauna are true indicators of chronological relations of fauna.

1. The lowest is a sandstone resting uncomformably on Azoic Carb[oniferous], Jur[assic] or other beds.
2. Dark shales, clays. Benton.
3. This is overlaid by gray white yellow chalk or calcareous marl–Niobrara.
4 and 5. Laminated shaly clays and sandy beds [Fort Pierre and Fox Hills].
Then Tertiary lacustrine[187]

The vertebrate paleontology shows the break to have occurred higher up in the series.

No. I. Dak[otah], 2500 [feet], according to Cope no vertebrates (Our Dinosaurs).
II. Benton, *Ostrea conjesta, Inoceramus prob.* Four vertebrates: *Lamia* (shark), *Pelecorapis* (a flying fish), *Apsopelin* (a similar fish), *Hyposaurus* (a crocodile).
III. Niobrara.[188]

## Trip to South Park, April 7, 1878[189]

I left Bear Creek to take a trip to South Park with a view to restoring health which had been a little run down. Passing up Bear Creek I noticed that the little *Cinclus mexicanus* or water ouzel [also known as American dipper] had nearly completed its nest on a ledge of rocks above a waterfall in the cañon. I had watched with interest the commencement of their arrangements for housekeeping as I passed the spot on March 23. The birds were then laying a foundation of mud in a little hollow in the rock, where they built last year. They brought the mud in their beaks and seemed to trample and knead it down with their feet, both birds working together to plaster a foundation for their future home. Now as I passed their work was well under way, the round penthouse was nearly built—nothing seemed to remain except to finish the door way which was a little too large. Both birds were working busily.

A few weeks after, on my return hoping to secure both nest and eggs for a lady, I rode to the spot with Mr. Allen.[190] May 30 the stream had risen to a flood in my absence owing to the rapid melting of the mountain snows on the summit of the ranges by the warm spring sun. The canyon was blooming with all kinds of beautiful flowers such as the white roses of the *Rubus deliciosus* [the Boulder raspberry, also known as the mountain thimbleberry] a bush covered with a mass of beautiful flowers and on the side of the stream just below the nest the elegant purple of bells of the *Dodecatheon* [*(pulchellum)* western shootingstar] were rising above the soft green moss and hanging pensively over the boiling stream. And a delicious fragrance filled the whole canyon, here and there on little beaches of river sand clouds of butterflies were basking, opening and shutting their wings as if in intense enjoyment of the warmth. Amongst them were the beautiful large yellow Machaon's kings[191] amongst the smaller blues and skippers that flitted about in joyous swarms whilst every now and then one of these beautiful Machaons would glide down the creek a few feet above the stream on easy wing dipping here and there to patronize a flower and then flying on again as if rejoicing merely in the power of floating flight.

On arriving at the waterfall near a little bridge, the stream and waterfall presented a formidable appearance and a serious obstacle in the way of getting at the nest! But naturalists are not to be daunted. I told Mr. A. to keep sentry on the bank for people going up and down whilst I climbed along under the rocks beneath the overhanging precipice till I could go no further. The nest lay on the rock only a few feet beyond me but to reach it it was necessary to wade into the boiling stream which threatened to carry me off my legs down over the waterfall into a perfect maelstrom of rushing water. However I resolved not give up without an effort. Probing the depth of the water with a pole, I found it would reach up to my middle. I took off my trowsers and stept in. The water was decidedly cool coming direct from the melting snows but I succeeded in getting on to a slippery rock from which I could just reach the nest above me. It came off the rock quite easily and without any break so compactly was it built. And I carried it with difficulty to the ledge unfortunately dropping out one of the young ones into the water. I was quite sorry to find it was full of young ones three or four in number. The old birds kept flying around me in great alarm yet singing all the while as is their wont. I debated for a moment whether I should kill the young ones and take this nest which I had so long coveted away, but neither science nor the lady could get the better of more humane feelings and replacing the nest on a ledge a little distance from its original situation we waited to see whether the old birds would return. This they presently did, flew to it in great anxiety picked off little pieces and dropped them in water as if to assure themselves it was really their own home. They flew to the old locality, examined it carefully and sat for some time meditatively as if considering the insoluble problem of so sudden a house moving. The male remained for a long time in this spot but the mother moved by her maternal anxiety hastened to the old nest to see if her children were all safe and presently commenced feeding them as usual. Then both took this duty upon them carrying food, apparently insects, in their mouths tapped and broke them on the rock with their beak before giving them to their youngsters. And sometimes the male would take the food from the female's mouth and give it to the impatient nestlings. As they appeared reconciled to their change of situation we left them after I had made a rough sketch of the locality, intending to return when family matters were settled and secure the graceful fabric which appeared to be made of moss and grass etc woven into a penthouse somewhat like a wren's nest. The bottom lined with clay where it was attached to the rock.

## Burning of Mathews and Jarvis Halls

This was in May but to return to our April trip; I met with no adventures till I reached a little roadside post office kept by some friends. Here in a *Tribune* [192] lying on the table I saw that Mathews Hall was burnt and with it my room and all my pictures, letters and furniture. This was a great shock to me but I afterwards learnt with relief that many of my things had been saved by the hook and ladder company although the room was blazing all around them and some of the men were much intimidated by a lot of cartridges I had in a box catching fire and going off in a promiscuous manner. My natural history collection of stuffed animals was also saved by the efforts of Mr. Everett, a naturalist friend, [a banker in Golden] who mounted guard over them with one of the sabres in the armory protecting them from people who were disposed to handle them much to the detriment of heads and necks of the stuffed birds. Many of which notwithstanding were reduced to headless masses of feathers. The library of the hall was entirely burnt. The fire was evidently the work of an incendiary and was not detected till had half consumed the building and what was saved was saved at great risk.

The following day through driving snow I reached the edge of the park and stopped at Mr. Bluebacker's hotel [located near Kenosha Pass in Park County]. The next morning was fine but I had not ridden far across the basin when I noticed some curious wreathes of white cloud blowing rapidly and chasing one another like phantoms across the snow of the basin issuing from one of the cañons of Silver Heels Mountain.[193] Sometimes they reminded me of a number of cannon going off from a battle in the mountains.

My curiosity as to this phenomenon was soon gratified by one of these phantoms rapidly approaching me and in a moment I was in a snow storm of snow blown from the mountains by the bitterest and most violent winds: storm followed storm in close succession till I was nearly frozen and hastened as fast as possible to gain a deserted shanty where I found an ex tempore blanket made of gunnysacks and used as a partition to divide the limited shanty in two. I tore this up and bound it tightly around my feet and legs and rode on again to another distant log cabin, where I was hospitably received by a coal miner and his family and it was good [and] warm at their stove and a pipe and a cup of coffee soon got us all to rights. Noticing a shock of hair hanging up against one of the beams of the cabin very much resembling an Indian scalp and thinking it might be such an interesting trophy, I ventured to ask what it was when Mrs. A. came out of the kitchen to inform me that was her back hair. I apologized of course. We spent quite a pleasant day and an agreeable chat about geological matters and the relations of the coal etc. in the park.[194]

Next day was as lovely as the preceding had been unpleasant and I had a delightful ride over the prairie turf to Mr. A's. Here I found all well and on

Sunday being Easter day had service with the children and christened the baby.[195]

The following week was spent in hunting ducks on the ponds but with little success as they were very wary. I succeeded however one day in crawling on my stomach like a reptile up to a lot of little teal that were feeding on the edge of one of the ponds and slaughtered six of them at a shot. I also killed two or three gray sandpipers or tip-ups[196] as they call them from their peculiar habit of tipping up or bowing especially when alarmed their whole body working to and fro as on a pivot formed by their long legs. These birds had long straight beaks, long legs and a grey plover like body about the size of an ordinary golden plover. They were in flocks of five or six and uttered loud cries when disturbed. I also put up and shot a brace of snipes. These lay quite as close as our English jacksnipe but not unfrequently I could see them sitting on the edge of the ooze near the pond which one never can in the case of the English snipes [the common snipe, *Gallinago gallinago*]. The size of the bird was between that of the English snipe and jacksnipe and the cry very much similar. They were comparatively tame and after being disturbed or even fired at would frequently alight within a few yards of the spot whence they were started. One afternoon I noticed a large bird about the size of a curlew wading in the pond and when I approached he rose with a most wild cry. I could not get a shot at it though I pursued it for several days from pond to pond a long whose edges it would wade and feed with most industrious rapidity. I believe from descriptions the bird was an avocet. I afterwards shot two birds in his company which I took to be female avocets. They were a little smaller than the ordinary curlew. Their beaks were long and slightly curved upwards. The breast was a light reddish brown and barred like a woodcock and the plumage of the back and wings resembled that of a plover or snipe. Their cry was very similar as to that of the male avocet who rose at the other end of the pond when I fired. The following is the description:

May 6, 1878, Avocets (More Probably Godwits) Beak four inches slightly curved up; length of body from head to tip of tail fourteen inches. Head greyish. Length of legs five inches. Tibia two inches, tarsus three inches. Longish middle claw one and half inches. Legs black. Beak lower mandible reddish black towards tip. Upper mandible longer by one eight inch than lower. Color of back like a plover greyish mottled black and light brown barred. Breast light reddish brown barred like a woodcock horizontally increasely towards neck. Primaries of wings reddish brown tipped with black shading off into reddish brown. Secondaries black tipped with brown. Scapularies brown and reddish. Length of primary ten inches to shoulder. Shoulder to body five inches.

Grey gulls and geese frequented the ponds but I did not kill any although I fired twice at the former with ball. One afternoon seeing a herd of antelope on the adjacent hill not far from the fence I took my rifle and crept towards them by a long detour so as to get the wind blowing from them to me. They were lying in a little hollow on the bare hill side and I crawled for long time on my hands and knees over the drift pebbles covering the turf and at last got to within fifty yards of them but I remained for twenty minutes without daring to raise my head. The herd were all lying down "nooning it" in the hollow. I could see their horns distinctly but as I saw dare not rise to get a shot. At last one got up to stretch himself. I could have easily shot him had it not been that I was anxious to get two in line but before I could accomplish this he saw me and I started to my feet with buck fever and as the herd all arose together fired indiscriminately into the pack and so missed them all. The graceful creatures cantered gaily over the hill and down its slope below me in a close column with the dust flying behind them as they passed I gave them another shot but the ball fell short and I missed them again, and they were soon out of sight flying on the wings of the wind.

On Sunday [May 12] I had engaged to preach at Alma and rode over on Saturday welcomed by Mr. M. They gave me the use of the Methodist church and my services were well attended.[197] On Monday Dr. ___[198] and myself climbed Mt. Lincoln. The weather was threatening and we had a hard climb over snow and ice meeting with a snowstorm half way to the Russia mine where we halted to dine. This mine is at an altitude of about 13,500 feet and the outside of it was completely banked with snow: The cook gave us a good dinner as usual and whilst Dr. went out [and] entertained me with the following stories of mountain rats

## Mountain Rats *Neotoma Cinerea*[199]

[end of journal][200]

March 19. 1880

I left Como for Golden
The strata along the U P R R for miles appears to be cretaceous especially Cret groups No 2 & 3 a great deal of dark grey shale is exposed in the cuttings The strata are upheaved into hogbacks [sketch] dipping about 25° & towards the south
At Miser is a deep RR cut showing large Cannon ball concretions in Cret No 3
At Wyoming station the dips is toward the North
As we approach Laramie we see Jellum Mt to the West connected with the main range dipping North
There is probably a wide Synclinal between Jellum Mt (Jurassic) and the Como hogbacks

[sketch: Jellum Mt — Como]

The sedimentary strata between the Laramie hills and the main range seems to have been wrinkled into several minor folds by lateral tangential forces exerted from the two ranges. The cretaceous is thus brought to light at various places and at Wyoming probably the Laramie Coal series

[sketch: Laramie hills — Main range]

The scenery along the track is the monotonous rolling prairie covered with snow with here and there a herd of antelope or some cattle. Every now and then we pass through snow sheds. Miser the short for misery is a lonely miserable spot for a station another station is Lookout Cooper lake lies in a depression probably a fold.

1879–1880

# Journal of Explorations for Saurians and Fossil Remains in Wyoming

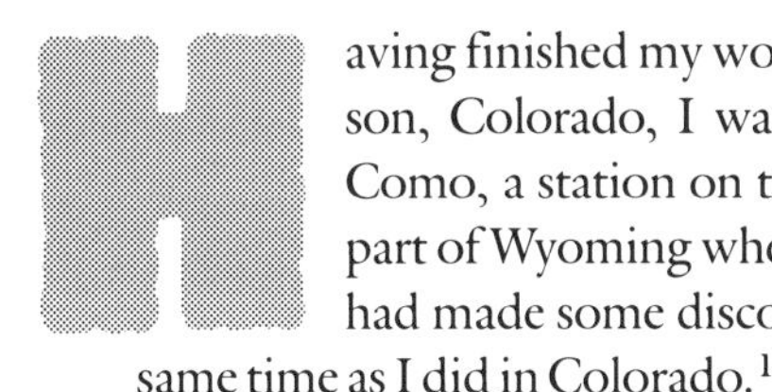

aving finished my work of exploration for Dinosaurs at Morrison, Colorado, I was directed by Professor Marsh to go to Como, a station on the Union Pacific railway in the southern part of Wyoming where a man named [William Harlow] Reed had made some discoveries of Dinosaurian remains about the same time as I did in Colorado.[1]

## Notes on the Road

May 14, 1879 I left Golden City May 14 taking the train for Wyoming.[2] We passed under the Table Mountains with their wall of basalt pillars like those of Fingal's Cave[3] past also the dyke filling the great fissure from which the lava once flowed over the surrounding country then out on to the open prairie robed in its tender spring green dotted over with herds of cattle. On we went past baby towns of yellow prairie buildings of only a year's growth. On past the great Triassic red sandstone cliffs rising up from the green prairie grass to a height of 3000 feet climbing the slopes of the Great Rockies whose range stretched in full view for a distance of 150 miles parallel with us on the west with white snowy peaks rising cool and cloudlike above the darker purple of the foothills clothed with pine forests and cloven by many a deep canyon.

Near Fort Collins the train passed through a cut in a long ridge of massive yellow sandstone full of large cannon ball concretions containing in their center great numbers of fossil shells of the Upper Cretaceous or

Fox Hills group such as big clams *(Inocerami)* ammonites, scaphites etc.[4] After this a monotonous prairie ride for about one hundred miles till we reached Hazard station where we made connection with the Union Pacific railway.

After leaving Cheyenne, we passed from the monotonous rolling prairie into a more broken bluff country dotted with fir trees and many fine masses of red granite spires and the continuation of the Rocky Mountain range. One town, we passed was a fort and barracks of the U.S. Army with soldiers walking about on the quadrangle in blue uniforms. This suggested a change in favor of civilization in contrast to the wild country we were entering.[5]

At Laramie City, Bishop Spaulding [Episcopal bishop of Colorado and Wyoming] boarded the train and accompanied me as far as Como on his way to a distant Indian station.[6]

## Como Station

About 8 o'clock PM we reached Como station, a lonely spot on the Union Pacific. The station consisted of a red building, a tank like a huge coffee pot and a small red section house for boarding the men at work on the track.

Bidding the Bishop good bye I jumped onto the platform and walked into the station house room. There were half a dozen men smoking and playing cards around the stove. Amongst them was a tall swarthy complexioned man with a handsome face dressed in a full suit of buckskin ornamented with fringes after the frontier fashion.

The broad brimmed white felt hat on his head had the front rim thrown up like a vizor and perforated by a gun-wad punch. He was standing with his back against the stove, the most striking and commanding figure of the group looking the ideal of a frank, freehearted frontier hunter.

I asked, "Is Mr. Reed here?" and he answered, "That is my name" and gave me a very hearty welcome. He had been expecting me for some days. The stationmaster gave me a comfortable bed in the station and I slept soundly after my journey.

May 15 After breakfast Reed and I started out to explore the neighborhood.

Behind the station to the south was a high bluff rising about 600 feet above the prairie, the top crested with sandstones of the Dakotah group of the Cretaceous. The face of the bluff is composed of ashen gray and variegated red and purple clays and shales with some them layers of sandstone.

This soft material is eroded here and there by rain and riverlets into many little channels and groovings and at wider intervals by deeper ravines. The upper portion of the bluff consists of these ashen and greenish strata passing down into a belt of salmon red and purple clays and shales. The reddish portion is banded with variegated belts and presented a handsome appearance from a distance.

The line of excavations in search of dinosaurs was a few feet above the variegated belt between it and the cap of Dakotah sandstones.

Reed and I took our picks and a rifle and game bag and began climbing the bluff to examine the different excavations that had been made at intervals along the bluff. These had been opened during the past winter and showed by numerous fragments of bones exposed that this bluff was very rich in saurian remains, far more so than those of Morrison and Cañon City in Colorado.[7]

The formations were very similar to those at Morrison but much better exposed, and the line of strata containing the bones could be followed with ease over a very long distance for six or seven miles at least.

## Quarry No. 3

The first quarry we came to was called No. 3, the quarries being numbered for convenience. It was about two miles east from the station along the bluffs. Several bones had been already taken out and despatched to Yale, amongst them a skull and jaws with herbivorous teeth in them.[8] We uncovered a mass of bones lying in the center of the excavation very friable and partially decomposed by the proximity of a small spring of water. They were mostly vertebrae and ribs in a tangled mass. Some of the small vertebrae were jet black and deeply indented on the sides and may be those of a carnivorous Dinosaur.[9]

## Our Camp

In the evening we climbed to the top of the bluff and looked down on the rolling prairie to the south to which the back of the bluff sloped gently at an angle of about 20 to 30°. We saw a few antelope at which Reed fired but without success.

At the foot of the Dakotah ridge was a little ravine cut through the dark gray shales of the Fort Benton group of the Cretaceous. In this ravine Reed had pitched his tent amongst the thick and tall sagebrush near a spring of drinkable but rather alkaline water. The hindquarters of an antelope were hanging from the ridgepole of the tent from which Reed cut a

William Harlow Reed, about 1910. Courtesy of the American Heritage Center, University of Wyoming.

## William Harlow Reed

William Harlow Reed was one of the greatest fossil finders of the American West during the nineteenth century, having codiscovered dinosaurs at Como Bluff in 1877. Born on June 9, 1848, at Hartford, Connecticut, his family moved steadily westward during the 1850s, arriving in Jamestown, Michigan, in 1863. During the Civil War, Reed joined the Union Army. After the war, he worked on the Union Pacific Railroad.

Reed's first job on the Union Pacific was shoveling snow off the track. He soon turned to providing game for railroad work crews much in the manner that Buffalo Bill Cody did at this time. In 1870 Reed returned to Michigan and married Florence Bovee and then began homesteading in Nebraska. His plans were abruptly shattered when his wife died in 1871 during the birth of their son, Oscar. Reed then apparently spent several years working as a scout, hunter, and guide in the American West. In 1874 he again worked for the Union Pacific Railroad and by 1877 was the section boss at Como Station in Wyoming.

There he discovered the fossil remains of a creature of such size that he and the stationmaster, William Edward Carlin, contacted O. C. Marsh regarding their find. Thus began Reed's association with Marsh, which lasted, to one degree or another, until Marsh's death in 1899. Reed was responsible for numerous discoveries during these years. His relationship with Marsh was at times strained by the lack of prompt payment of funds and Marsh's hiring of other collectors to work independently of Reed at Como.

In 1880 Reed married Anne E. Clark, of Milford Center, Ohio, whom he had met through correspondence related to an advertisement he had placed in the Detroit Free Press, offering to exchange fossils and Indian artifacts for books, particularly by Dickens and Thackeray or about Hebrew philosophy. Reed left Marsh's employment in 1883 to try sheep ranching in the Shirley River basin. However, he was not successful and returned to collecting fossils for Marsh and others, guiding hunting parties, and doing odd jobs during the next decade.

In 1897 Reed accepted a position at the University of Wyoming as assistant geologist and curator of its museum. Besides collecting more than eighty tons of specimens for the university, Reed continued to collect independently. In 1899 he resigned his position in order to collect for the Carnegie Museum. A New York newspaper article had described the discovery of the skeleton of an incredible creature, eleven stories high and 130 feet long by one "Bill Reeder." Andrew Carnegie instructed his museum director to obtain this specimen. The newspaper account was grossly inaccurate, but that summer Reed did help find a species of diplodocus *(D. carnegii)*. The mounted skeleton of this diplodocus was cast and copies sent to museums all over the world.

In 1900 Reed left the Carnegie Museum and began three years' employment for the American Museum of Natural History, during which time he located a new quarry and also worked at the Bone Cabin Quarry. In 1904 he was hired as assistant geology professor and museum curator at the University of Wyoming. He held this position until his death on April 24, 1915. Overall, Reed was a remarkable, self-taught, pioneering American paleontologist and frontiersman—an American original.

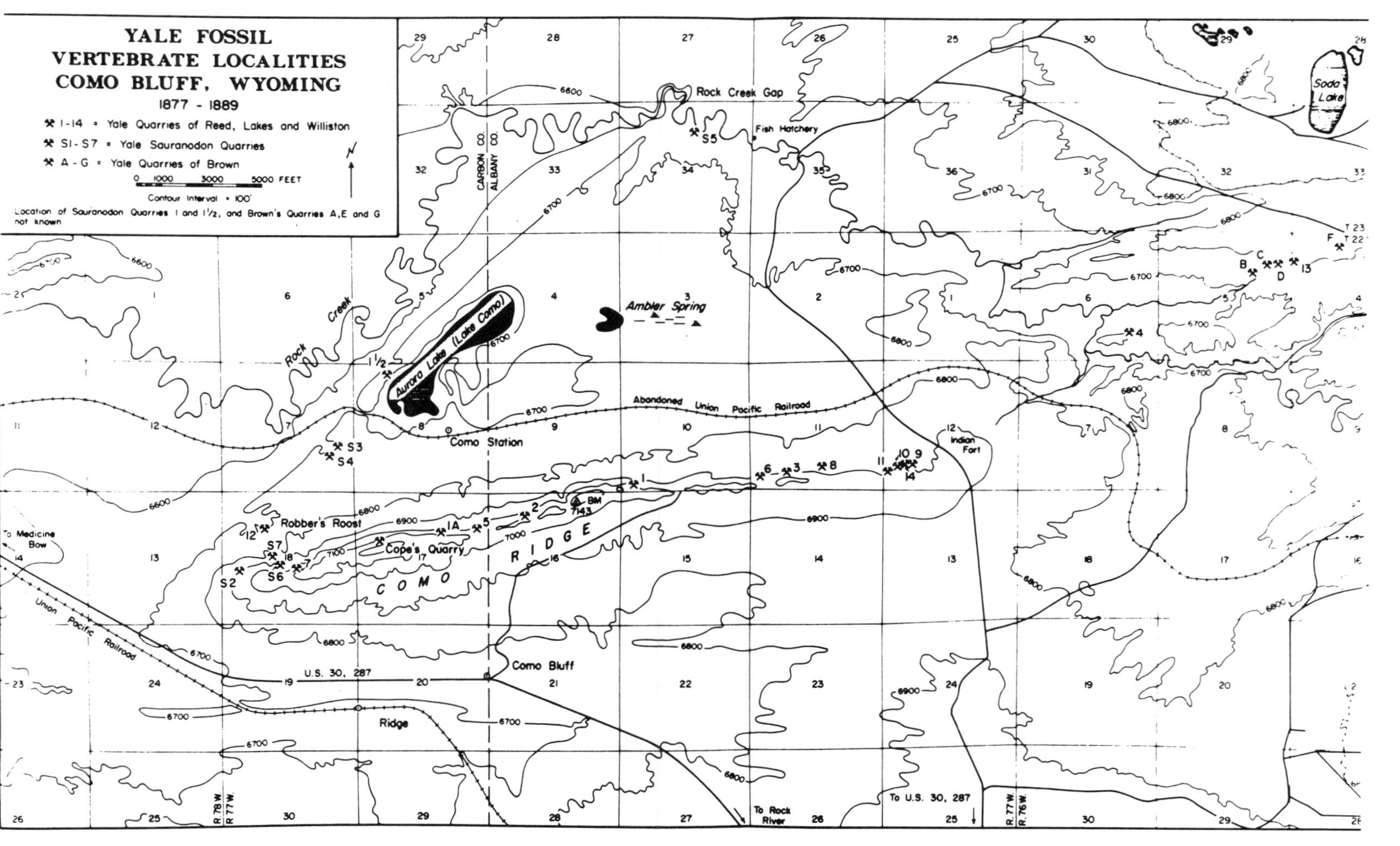


Yale fossil vertebrate localities, Como Bluff, Wyoming, 1877–89. Courtesy of the Yale University Press.

steak for our supper. A couple of wooden bunks or rude bedsteads formed the tent furniture.

After a hearty supper on antelope steak and camp bread, we sat outside smoking our pipes and becoming better acquainted, more communicative and confidential to one another. As we watched the stars climb up behind the bluffs of gray shale on the tops of which stunted weather-beaten fir trees grew and sent down over the bank extraordinarily long serpent-like roots, very suggestive of the dragons sleeping in the stone beneath them, we made up our bed at the floor of the tent and were soon sound asleep.

## Discovery of *Ichthyosaurus (Sauranodon)*

May 16 We were up soon after sunrise. It was a fine fresh morning. The sun was rising behind the bluff and flooding the rolling prairie with golden light. We washed at the spring and then had breakfast which was much the same as our supper. We started in a westerly direction along the ridge through a thick jungle of sage brush with its matted tangled roots reminding one somewhat of the English gorse and furze. Our destination was to examine, and if possible, get out the bones of an ichthyosaurus (Fish lizard)[10] which Reed had seen protruding from a concretion of limestone on the side of the cliff on one of his hunting rambles. He said it was situated in a bed below the *Atlantosaurus* horizon[11] which was full of belemnites, ammonites and small shells, evidently in marine strata of the Lower Jurassic and probably in a horizon corresponding to the Lias of England which is so rich in ichthyosaurian remains.[12]

After an hour walking we reached a point in the ridge which looked down on the locality we were aiming at. As we descended the steep face of the bluff from the Dakotah ridge we found immediately under those characteristic Cretaceous sandstones a bed about five feet thick of dark black sandy shale evidently carbonaceous and containing in it many well preserved fragments of silicified fossil wood. Below this were dark olive green concretions just above the *Atlantosaurus* horizon. As we climbed down over the soft *Atlantosaurus* beds we noticed on the opposite side of a little ravine some large fallen slabs of sandstone beautifully indented with fossil ripplemarks.[13]

Descending thus the cliff at about 300 or 400 feet below the Dakotah cap we reached a terrace of soft massive cream or sulphur yellow sandstone much jointed and eroded into forms something like basaltic pillars. This Mr. Reed considers as the cap or top of the Lower Jurassic or marine (Liassic beds). Below this is a bed about 100 feet thick of fine drab shale on the surface of which lay numerous fragments of belemnites *(Belemnites densus)* from three to six inches long and resembling the ends of half smoked cigars; here and there was a hard limestone concretion containing some large ammonites eight or ten inches diameter and also impression of a smaller species. The calcareous [containing calcium carbonate, calcium, or limestone] slabs and shales contained or were at times entirely composed of quantities of small bivalve shells forming locally a shell limestone. At the base of this calcareous and shaley bed is an exceedingly massive cream colored sandstone showing wonderful examples of cross bedding and sometimes eroded into curious monuments. This bed, which averages 100 feet thick, is usually considered in Colorado as forming either the base of the Jurassic or the extreme top of the Trias. The Red Trias conglomerate follow immediately upon this as at Morrison, Colorado.

It was some time before we could find the exact spot where Reed had noticed the bones. At last we found it: a large calcareous concretion with some ribs and other bones sticking on it. It lay attached to the cliff but exposed by a little gutter formed by erosion. On following this gutter down the hill, we picked up numerous fragments of vertebrae and ribs; the peculiar disk-like form of the former at once showing it to unmistakably belong to the *Ichthyosaurus* family. The largest of these vertebrae that we found was about four inches in diameter and so thin in the center that finger and thumb almost met when pinching it. The shape and biconcave character of the vertebra showed that it did not belong to any land animal but rather to a very large fish or marine saurian. On removing [a] portion of the covering of the concretion we were disappointed in not finding part of the skull, great saucer eyes, or teeth as we had hoped from the shape of the concretion, but only a few ribs and vertebrae. Nor could we on this occasion by a few feet of digging find traces of the rest of the skeleton extending into the bank. We determined however on a future occasion to make a deeper excavation and more thorough search. We had succeeded in finding among the bushes and debris quite a number of good fragments as much as we could carry down to the railway track about a mile off, where we concealed them temporarily till Reed could get a chance of borrowing the handcar and fetching them up to the station.

## Train Wreckers[14]

Whilst resting on the rocks above the railroad Reed pointed out a little trestle bridge where a party of desperadoes or "road agents" had attempted

to wreck a passenger train a short time before my arriv
plunge them down an embankment twenty feet deep wh
been certain death to many and then rush in and plunde
confusion of the disaster. Fortunately the rails which they had displa
for this purposed were discovered by a train walker in time and the plot prevented. He showed me a ravine where the desperadoes camped. The sheriff named Widdowfield and a deputy with western pluck and temerity undertook to follow up the gang and were both shot and killed from ambush. One of the gang named Dutch Charlie was afterwards captured and confessed the crime. As he was being brought to Carbon to be tried, the miners entered the train, took out the prisoner, hung him to a telegraph pole, and riddled his body with bullets. Such incidents are common enough in this wild lawless country which is a paradise for the desperadoes and offscum of civilization. Yesterday the train went past with an armed body of soldiers to look after "road agents" and Reed whiles away the hours around the camp fire with stories of desperadoes and shooting affrays that he has experienced.

We went to the station to fetch my valise and things to carry them to camp. A tedious work dragging them up the steep face of the bluffs in the darkness. My valise broke loose in the operation and many of its contents fell out on the rocks. As it was too dark to look for what we had dropped, we set a mark on the place and reached camp for supper and bed.

May 17 We had intended spending the morning in rigging up a bunk and bed for me in the tent and making other arrangements for our comfort, but on arriving at the station the railroad hands begged us to come and pitch our camp near the station for the sake of company. They goodnaturedly volunteered to go over to our camp the next morning and bring over our things. So that evening on returning to our bivouac, I made a sketch of our first camp in the wilds and the following morning the "boys" appeared in force.

May 18[15] Next morning the hands of the railroad appeared soon after breakfast volunteering to carry our goods and camp outfit to a more convenient spot for water and mail near the railroad station. Our tent was soon struck the party filing over the ridge. One with our camp stove and another with our tent over his back and others following up the rear with various camp articles presented a comical appearance reminding one of a party of Ute squaws moving their teepees along the trail. As soon as we reached the

Frank James, about 1905. Courtesy of the Kansas State Historical Society.

## The Train Wreckers

The attempted wrecking of the Union Pacific train that Lakes mentions several times in this journal was an important incident in the history of outlaws in the Old West. Its place in western history is due not only to the notoriety of its members—Frank James, brother of Jesse James, likely participated in it—but also to the infamy of wrecking a train and ambushing a posse as well as the gruesome deaths some of the outlaws met. Lakes was mistaken concerning some of the details, such as when the attempted wrecking the train occurred.

In 1878 a gang led by "Big Nose George" Parrott (also known as George Francis Warden), with Frank James (using the alias McKinney), "Dutch Charlie" Bates, Frank Towle, Jack Campbell, Joe Manuse (possibly), and one other plotted a major train robbery of the Union Pacific Westbound Express no. 3. In order to stop the train, they employed a method used by the James gang in Iowa in 1873: derailing the train by removing a rail of the track. After breaking into a toolshed on Saturday evening, August 17, 1878, they removed a rail about a mile from Como station where the bridge crosses Rock Creek. According to a newspaper account, about 8 P.M. that evening a gang of section men noticed that a rail had been completely removed together with its spikes, and they repaired the damage before the train arrived. Another account of the incident states that a lone section foreman noticed that the rail was loosened, not removed, and he continued along in his handcar without appearing to notice the damage while the bandits debated killing him. In any case, the train was warned and the wreck prevented.

A posse was soon organized but split off in a number of directions to search for the gang or wait for reinforcements. Two members, Deputy Sheriff Robert Widdowfield of Carbon and Union Pacific detective Henry "Tip" Vincent of Rawlins, proceeded south toward Elk Mountain. They tracked the outlaw band up Big Canyon by Rattlesnake Creek, where they were ambushed. Some twenty shots were fired and both Widdowfield and Vincent were killed. Jack Campbell took Widdowfield's boots and "Dutch Charlie" the best saddle. The gang then continued in haste south toward the Colorado's North Park.

The gang broke up, but a number of its members came to notorious ends. Lakes's account of the death of "Dutch Charlie" Bates is close to the facts. Bates was captured on January 1, 1879, at Green River, taken first to Cheyenne, and then on the evening of January 5 transferred by train to Carbon to await trial. As the Union Pacific train pulled into Carbon City, a party of twenty-five to thirty masked men, identified by a newspaper as miners and citizens, boarded the train, seized Bates, and dragged him off the train. The prisoner was stood on a barrel and a noose strung up over an adjacent telegraph pole. Bates admitted his part in the deaths of Widdowfield and Vincent and begged to be shot instead of hung. His body hung on the pole until the afternoon of the next day as a warning to malefactors. The *Cheyenne Daily Leader* commented that the train was delayed by only thirteen minutes by the incident. The *Cheyenne Daily Sun* titled its article "Lynching Last Night: 'Dutch Charlie' Takes His Last Dance in a Hemp Necktie, with a Telegraph Pole for a Partner" and warmly approved of this method of dealing with road agents and cattle thieves.

Frank Towle was killed on September 13, 1878, while attempting to rob a Cheyenne-bound stagecoach from the Black Hills. A gang of bandits stopped the stage and demanded the mail sacks and money of individuals but two mounted guards engaged the robbers

Continued on page 94

in a gun battle in which Towle was killed by Boone May. That December, another member of this stage robbery, John Irwin, confessed to Sheriff N. K. Boswell of Laramie his role in the robbery and told of the death and gravesite of Towle. The sheriff told Boone May, who disinterred Towles's body, cut off its head, and tried to collect rewards in Laramie and Cheyenne. Boone failed in this grisly business and the skull was discarded.

"Big Nose George" Parrott and Jack Campbell were both arrested in Miles City, Montana, in July 1880. Campbell escaped on the way back to Rawlins, but Parrott met with much the same reception as "Dutch Charlie" when the Union Pacific train arrived in Carbon City on August 14, 1880. A gang of masked men stopped the train and used a sledgehammer to break the chair in which Parrott was chained. They dragged him from the train and with a rope hoisted him the neck until he confessed. This time the prisoner was returned to the sheriff. In December, Parrott was sentenced to be hanged. On March 22, 1881, he attempted to escape by overpowering Sheriff Rankin but was stopped by Mrs. Rankin, who locked the jail cell and pointed a revolver at Parrott. That night a mob lynched Parrott. A local doctor, John Osborne of Rawlins, requested the body for scientific purposes. He made a death mask of Parrott, cut his skull open, examined his brain and had a pair of slippers made out of his skin. His skullcap was presented to Osborne's medical associate, Dr. Lillian Heath. The death mask and slippers were on display in a bank in Rawlins for many years. Dr. Osborne was elected governor of Wyoming in 1893.

Joe Manuse was tried and found not guilty of the murders of Vincent and Widdowfield by a district court in Rawlins on September 10, 1879. Frank James was never arrested for this attempted robbery. He died in 1915.

railroad the tent was pitched opposite the station and immediately it was filled with our railroaders: Irish navvies [laborers] and station men stretched out at full length smoking and chatting. They were goodnatured fellows though their language was not the choicest. In fact the profanity of this region strikes one at first as something awful until by its daily recurrence we become, as it were, used to it. Our little tent is filled nearly every day with such a crowd and especially Sundays when there is no work going on. They are our only society and are all wonderfully obliging and willing to help us. Sometimes however I could wish for a little privacy in which to read and write. My sketch book is a great source of amusement and interest to them.[16]

After dinner, Reed and I and a man named Kennedy[17] walked over to the other side of the lake and explored its banks. Lake Como which gives its name to the railroad station is a shallow piece of water about one mile long by a half mile broad. It lies in a basin scooped out of the top of a low arch or anticlinal of the red Triassic sandstones. It is famous for a species of water lizard allied to the Mexican axolotl called *Siredon amblystoma* which passes through many interesting transformations ere it reaches its adult condition. In one state it resembles a full size trout about a foot long with curious fringed gills hanging from the head. This is, so to speak, its tadpole state. Legs then begin to appear and finally it drops a good deal of its tail and emerges from the water as a crawling newt or water lizard.[18] The water of the lake is strongly alkaline and unpalatable. It is a great resort for wild fowl, many of whom build in the reeds along the banks. There were a great number of the eared grebe *(Podiceps auritus)* [horned grebe] swimming about together with coots, ducks, and a few gulls (probably Bonaparte's gull). We noticed a pretty specimen of a tern dipping gracefully into the water like a swallow. Its wings appeared to be of an ashen gray color and its body blackish.

Near one of the sandstone ridges forming the North bank of the lake, we came upon a quarry of saurian bones but slightly opened: A medium sized sacrum of a dinosaur lay exposed on the surface.[19]

Along the banks some hedges of brush had been planted by the hunters as cover to shoot at the wild fowl which swarm here in the spring and autumn.

On our return to camp, we found that Carlin[20] had killed a black-tailed deer and brought it in.

## Eagles Nest; Arranging Camp

May 19 Reed and I went to the *Ichthyosaurus* quarry[21] to finish packing up the bones we had found there.

I sketched the locality. Reed pointed out a golden eagle's nest in the cliff above the *Atlantosaurus* beds. Last year he succeeded in shooting the pair and was let down by a rope to the nest but failed to find either eggs or young ones.

We returned to camp to dine on a haunch of venison which will keep us in provisions for some days—till Reed shoots an antelope.

We spent the afternoon in putting our camp in order laying down a carpet made of ripped gunny sacks [burlap bags] stitched together and pinned down into the turf with gigantic carpet tacks viz, railroad spikes, six inches long.

We made a table of boards and filled empty boxes into cupboards for our pantry; also made a rough bedstead and filled our bed ticking and pillows with hay from the marsh.

Then we dug a hole in the spring and sunk a barrel for a supply of fresh water. When all our arrangements were through the railroad hands pronounced it "The boss camp" and it certainly did look quite comfortable.

The men caught a small species of prairie squirrel *(Spermophilus richardsonii)* not unlike a small prairie dog but of a variety I had never seen before. Its holes were all around the camp and the creatures were very abundant; we constantly heard their shrill, bird-like whistle. The men caught the one by pouring water down its hole till the poor little creature was obliged to come snuffling and panting to the outlet and was easily caught.

May 20 We get up about half past five to breakfast and at seven the railroaders call for us on the handcar.[22] We have an exhilirating ride in the fresh morning air. The handcar propelled by six strong fellows flies over the rails of the railroad track making the rabbits and spermophiles dart from the bushes in every direction. Our detination was the quarry [no. 3] where we had lately begun to uncover some bones. After uncovering a large mass of clay the pick rang upon a monstrous limb bone four feet three inches long and nearly three feet in diameter at its broadest end. It proved to be a gigantic ilium with the pubis lying close to it. These bones appeared to be mostly vertebrae and ribs: three cervical, are connected with the hips and lumbar.[23]

On our way back to camp Reed showed me the skeleton of an elk he had killed some time ago. It was picked perfectly clean by magpies, ants, and beetles. As we were walking through the sagebrush, a dozen long necks suddenly appeared belonging to a flock of sage hens [sage grouse *(Centrocercus urophasianus)*]. Reed's rifle was at his shoulder in a moment and a fine cock bird was stretched on the ground. It is a handsome bird be-

*W. H. Reed and E. Kennedy Seated at Quarry No. 3.* Watercolor by Arthur Lakes. Courtesy of the Peabody Museum of Natural History, Yale University.

longing to the grouse family about the size of a large domestic fowl with gray and brown partridge-like feathers and a ruff of feathers around its neck. The feathers of the tail opens into a broad fan like a turkey's which it spreads out and struts about proudly for the admiration of the hens. No one however around camp will eat the flesh partly from prejudice and partly on account of its sage-like flavor.

At 8:30 P.M. the train brings the mail so eagerly expected in these lonely solitudes.

May 21 Went to the quarry on the hand car as usual and sketched the bones as they lay exposed in the quarry.[24] Reed gave us a treat of strawberry shortcake for supper the strawberries being canned. However they went down with great gusto. The hands came in after supper. I sketched one of them as a figure for my quarry.[25] Our night's rest was disturbed by the howling of the a small dog and by a chorus of maniacal laughter from a band of coyotes also by the constant railroad whistles. R took vengeance on the wretched cur next morning by a shot from his rifle.

May 22 Left camp early as usual on the handcar. We saw a herd of what afterwards proved to be antelope but by the mirage appeared so large as elk. Reed started in pursuit whilst I went to the quarry and found to my dismay a bank of clay had fallen in on the bones nearly demolishing some of them.

Reed appeared on the edge of the quarry with his hands covered with blood a proof that his hunt had been successful. His supposed herd of elk proved to be antelope of which he had shot a fine buck. He had cut off the hams of the hind quarters and also the head with its fine pair of horns.

After removing the fallen earth, we cleaned the bones of the saurian with our knives. I sketched the broken pubic ? bone and we packed it up first so as to get at the rest of the ilium which ran down beneath it for some feet into the clay. There were several small caudal vertebrae associated with the latter about three inches long. The ilium? [scapula] appears to be about four feet long, a rather shapeless mass of bone.

At lunch the coffee pot upset amid the anathemas of the party for it meant a dry dinner for thirsty men; no water being obtainable.

On top of this came another disaster. The head of John [George] Leech, an Englishman,[26] appeared over the dump and he handed a letter in pencil from Mr. Chase the stationmaster[27] saying that our tent was blown down and badly torn. We quit work immediately and repaired to camp to find our tent sprawling on the prairie and our papers and light material blowing over the country. Mr Chase helped us to sew it up again. The railroad hands helped us to pitch it again. We soon forgot our disaster over a steak of antelope fried for supper.

May 23 Blowing hard from the East: Heavy work beating up against the wind in our handcar.[28] Covered the bones with soil to prevent the danger of banks of earth falling on them. Packed some of them to the railroad track and flew home to rest at camp on the wings of the wind. Spent the afternoon in writing letters, making boxes for shipment, cooking, and putting camp in order for Sunday.

Reed proposed to go to Medicine Bow and suddenly appeared transformed from a buckskin hunter into a well dressed polished booted citizen.

He strikes me as a very fine fellow. After the supper by the camp fire he has told me his history. He had a common school education in the East. Went into business in a saw mill and settled on a farm where his wife died and his home was broken up. Broken hearted and reckless he took his rifle and went West careless, how or where he spent his time, sometimes as a hunter, prospector, guide, and till lately foreman on the railroad track, a splendid shot, a keen sportsman, a lover of nature, natural history and scenery, a wonderful power of close observation which with his occasional intercourse with stray naturalists and scientists had made a very fair naturalist of him. He had become too a fair practical geologist and had increased his knowledge in that science by the study of books lent him. A man of much force of will and character qualifying him as a leader, free-hearted and generous to a fault and honest and honorable. We get along first rate together, sleep in the same bunk, eat and drink and share all things in common except the prevailing western habit of profanity. He was the discoverer of the fossil bones in this region and has worked several years for Professor Marsh.

He has an endless fund of stories of the wildest character of his experiences in the parks and mountains amongst Indians, bears, desperadoes; stories which bear the character of truth. He is quite our leading spirit around here.

As for the rest of the "society" at the station there is a respectable young fellow and his wife who are very kind on watching over the safety of our tent and goods when we are away and helping us in many ways.[29] Then there is a young man and his brother who are bone digging for Professor Cope, Marsh's rival. He is rather an idle chap.[30] The section boss of the boarding house is an illgrained Irish man and his wife.[31] The foreman on the track is a good natured young fellow who does all kinds of good turns for us by carrying our bones and oneselves on the handcar [Edward Kennedy] and the rest are what in England we should call navvies. Taken it all together, though our society may be a little rough it is good hearted, free and easy and we throw in our lot together without stiffness.

As I am writing this, Reed the hunter is lying full length stretched smoking on the bed in the tent. Cope's men and three or four other railroad hands and the station master are also lying around on the grass smoking or trying to extract a tune from some vile musical instrument resembling a pandean pipe which with two Jews harps, Reed brought with him last night from Medicine Bow with great glee and triumph together with other things of a more substantial nature such as groceries, canned fruit, potatoes, an oil cloth for our table, a hair comb and other camp luxuries. The trains constantly passing and repassing keep us in mind of our being, after all within reach of civilization. It seems curious however from our wild life to look in on the luxurious cars and see well dressed gentlemen and ladies like phantoms we once knew passing by us from some other world. We have too a regular daily mail and can send to Denver for anything. Still I am often impressed with the intense wildness and solitude of the region. There is not even a farm within ten miles of us. An army of tramps are perpetually passing along the track; relics of Leadville and Black Hills excitements,[32] a class Reed detests with whole souled fervor. Once in a while the railroad kings pass down the track to inspect the road. One great treat is letters and papers when they come. I'm able to study by snatches but mainly scientific works relating to our business, and also I have abundant opportunity for noting the habits of birds, animals, insects, and collecting flowers. The region teems with fossils of the Jurassic and Cretaceous periods and what with books, stories, plenty of work muscular and mental and good health, the life is quite enjoyable; and so I have wound up my second week at Como.

May 24 Breakfasted at the section house, wrote letters home. The camp being quiet, spermophiles stole out of their holes and sat up on their haunches near the door of the tent eating and sometimes quarrelling over bacon rinds; at other times whistling or making raids into the tent. Reed returned from the village with two Jew's harps and a pandean pipe on which the boys began to play sacred melodies.

In the afternoon I strolled along the edge of the lake. Coots and grebes swarmed on its surface. The latter are conspicuous by their long necks and eared heads and by their constant cry resembling "coy-eet!" They swim fast and break the placid surface by a long wake or ripple after them. Their necks only are visible, their bodies lying so low in the water. The north shore of the lake is bounded by a steep ridge of gray Triassic sandstone and by the upturned Jurassic and Cretaceous beds. From the top of this ridge we would peep over and surprise the little sandpiper *(Tringa macularia)*[33] bobbing and bowing in the most comical manner on their delicate nervous legs. I saw a couple of cinnamon teal rather a rare bird with their copper tinted wings burnished by the setting sun. The plumage of the hen bird is light brownish gray more like a hen mallard. They were very tame and sported and played about not more than twenty yards from me. The wild ducks were more timid. On the shores grew a very sweet smelling white flower not unlike jessamine but growing on a little bush resembling the heath family *(Erica)*.[34] Dwarf varieties of the deep blue bells of the *Pentstemon*[35] and a bright scarlet paintbrush *(Castillera) [miniata]* grew on the banks. From the sagebrush numbers of bush and jack rabbits started as I passed through them. On reaching the farther end of the lake I noted numerous deer and elk tracks. Here there was an old raft and a rude boat that had been used by the duck hunters. The "kill deer" plovers ran along the beach with their plaintive cries and by feigning to be wounded proved that their nest and young were not far off.

I pushed the cranky boat out from the shore and tried to pole it to the other side but it leaked badly so I ran it into a little bay and landed among the reeds, the favorite nesting grounds of the grebes, ducks and wild fowl. A muskrat ran close to me along the bank. I reached camp, smoked and chatted and went to bed.

May 25 I spent the morning in sketching bones of No. 3 and the locality of the *Icthyosaurus (Sauranodon)* quarry. Being all alone in the tent I had as usual the company of the spermophiles. In the evening I strolled up the railroad track to meet the men coming back in the handcar. Carlin had killed an elk and they were bringing back the hams. After supper the story of the chase was recounted which started Reed and another hunter off on many yarns of their experiences in hunting.

May 27 Reed had found part of a saurian jaw with herbivorous teeth in it at Quarry No. 3.[36] So we went up on the handcar to the quarry to see it. There was about six inches of jaw rather decayed and portions apparently of the skull in a very friable condition attached to it with another larger fragment about a foot from [it], also perhaps part of the skull. We saw a couple of deer and a fawn near the quarry. The day spent in shovelling out dirt.

We had elk meat for supper which was tender and nice. We made a cabinet for geological specimens out of a dry goods box. Caught an axolotl [tiger salamander] in the railroad [water] tank. Reed told a story of how he found the old deserted camp of some camping party in North Park:[37] He found a broken fiddle of beautiful workmanship and some blankets. The campers must have been wealthy people and had probably been suddenly run down by Indians and either made a hasty retreat or else were massacred and never heard of. He also told about finding the skeleton of a man and a gun in a fissure of rock. He considered the man must have been frozen to death.[38]

May 28 Spent the morning sketching the bluffs from the north shore of the lake. Clouds of cliff swallows were flying about on the ledges or rock forming the bank of the lake or resting basking on the sand near the water.

The strata on the north shore belong to the same geological horizons and groups as those forming the bluffs south of the station, viz Jurassic and Cretaceous.

## Geology of Como

The strata of the north shore dips steeply at an angle of 50° and towards the north. The lake is hollowed out of portions of an anticlinal, thus:

Mainly out of the central arch of the anticlinal formed by the Red Beds or Trias.

Reed returned from Quarry No. 3 reporting the discovery of a large humerus bone.[39] Men discovered an enormous claw core of a carnivorous dinosaur much curved; sharp and long.[40] We find bones of herbivores and carnivores often mingled with teeth of both species. The carnivores are sabre shaped, the herbivores of blunter and more spoonshaped character [see facsimile, p. 000]. Walked out by the shore of the lake at sunset. There was a muskrat swimming in the rushes and a couple of ducks flew close in under the bank unconscious of my presence the female quacking lustily. I could have dropped a stone of them.[41] It was a wild scene with the glare of the sky reflected on the lake so far away and solitary on the desolate Wyoming prairie.

May 29 Walked with Reed to Quarry 3. We uncovered portions of a sacrum consisting of two united vertebrae *[allosaurus sacral]* lying beneath the humerus. Reed unfortunately tapped a small spring which filled part of the hole. Reed found three pretty small and rather constricted vertebrae a yard from the humerus.

In the evening arranged Reed's cabinet of fossils.

May 30 Shovelling earth.

May 31 On our way to the quarry we saw a herd of antelope. Reed started in pursuit whilst I went on to the quarry. Soon I heard him fire and saw him lying on the prairie firing shot after shot at the herd which were flying over the prairie. Puff after puff of white smoke came from his rifle and I could also see the dust rise where the bullets struck sometimes behind or in front of the flying herd. He was trying thus to corral them into a bunch. After twenty four shots he succeeded in wounding a fawn which dropped behind the herd and after six more shots he killed it.

On our way home on the handcar we saw a deer about five hundred yards from the railroad track. Reed was immediately in pursuit. The deer bounding with graceful leaps up the steep slope of the bluffs, was soon lost over the crest of the hill.

Reed pointed out to me a narrow but distinct trail about a foot wide and half a mile long which he said was made by jack rabbits. There was no sign of their feet but the grass grew in a defined line as distinct from the prairie turf.

## Discovery of Quarry No. 6

In the afternoon we measured the strata 500 yards west of No. 3 with a tapeline. The crest of the bluff is composed of a coarse conglomerate sandstone of the Dakotah Cretaceous at the base of which are silicified stumps of trees. Below this is a bed of black lignite shale, then 125 feet of drab clays and shales with beds of concretions.

About half way up this group we found a very small perfect limb bone apparently a femur six inches long, also a claw core one inch long and some small biconcave vertebrae.[42]

Below this were the series of variegated clays and shales, purple, red, and greenish traversed here and there by thin beds of shaley sandstone weathering brown which continues to the yellow sandstone capping the marine or lower Jurassic containing the *Ichthyosaurus* bones.

The men found the nest of a sage hen behind some railroad ties near the track. They brought in six eggs each about two inches long and of an olive greenish color profusely speckled with little brown spots. The nest was a very rude one consisting of a few blades of grass loosely woven together and some feathers lying in a little depression at the foot of a bunch of long grass.

June 1 Read *Vestiges of Creation*[43] to Reed in the tent.

June 2 Hired another man.[44] Watched a coyote sneaking among the bushes. Reed shot it. Then found small saurian bones 500 yards west of No. 3 where we made the section. They were small vertebrae apparently hollow with carbonate of lime in the center, also a curved sabre tooth about a inch long belonging to a carnivorous dinosaur. In the afternoon I sketched our camp. The last discovery Reed calls Quarry No.6. The remains appear to belong to a small carnivorous saurian.[45]

June 3 Sketched sacrum at No. 3. Two vertebrae each six inches long at No. 6. The men got out several small vertebrae each about an inch long. The processes attached to them seemed enormous. We also found several limb bones of this animal.[46]

## Professor Marsh's Visit to Camp

June 4 Professor Marsh whom we had been expecting arrived and breakfasted with the hands at the sectionhouse and soon made himself at home with the men.

After breakfast we started a large party of us on the handcar the "rubber car"[47] being attached to it for Marsh and my benefit. It was a lovely morning and Professor M was much amused at the innumerable spermophiles, rabbits and hares that kept leaping in every direction from the railway track as we sped along.

We stopped first at Quarry No. 3. Professor M, after examining carefully the bones devolved many of them to destruction as too imperfect or rotten for preservation. What we had taken for an ilium proved to be a scapula. The pinched vertebrae belonged to a carnivorous dinosaur and were caudals.[48]

Thence we went over to our new discovery at No. 6 the bones there are those of a crocodile with scutes or scales of armour deeply pitted with little holes. One bone he thought to be that of a pterodactyl, a small half inch jaw might belong to a small mammal. The teeth were those of a crocodile.[49]

From there we again took the handcar to Quarry No. 4 which lies on the side of a deep ravine hollowed out by Rock Creek, some four or five miles east of Como. This quarry had been worked by Reed previous to my arrival at Como. Several large bones were visible. Near this quarry we found a fossil shell resembling a snail or (helix).[50] It was in a concretion between the quarry and the overlying Dakota sandstone.

From a ravine close by a doe elk suddenly started from the bushes and ran up to the top of the ridge. Reed who happened to be on ahead did not see it at first till Marsh and I shouted to him. He wheeled round and fired directly. He caught sight of the beast striking him near the hip but not wounding him enough to stop him. Reed got another shot but without success. The elk gained the top and was lost to sight over the ridge. Reed followed him and tracked him for an hour by drips of blood.

The bones of this quarry were those of a sacrum probably belonging to *Morosaurus*.[51]

We returned to the handcar to lunch. The railroad hands and our party made quite a picnic sitting around the viands which were spread out on the rubber car.

After dinner we went to Quarry 1 whence Reed had formerly sent some very fine bones of *Morosaurus* to the Yale Museum.[52] There lying on our stomachs or on hands and knees we all hunted for small bones in a microscopic manner and were rewarded by finding what Marsh thought might be the toe bone of a pterodactyl or a bird.[53]

A few bones found in sandstone were also examined by Professor M and then we turned our attention to the *Ichthyosaurus* or *Sauranodon* horizon and quarries.

In these beds Marsh found amongst the ammonites and belemnites some fragments of a crinoid *(Pentaerinus asteriscus)* a typical Jurassic form. Marsh estimated the thickness of the Jurassic bed at about four hundred feet. We sat on the bluff watching a miniature whirlwind brawl along the railroad track [which] according to Marsh's watch [sped along] at one mile a minute. We watched it reach our tent and take the stovepipe off. Marsh examined some loose vertebrae of *Sauranodon*. He is anxious for us to find some paddle bones. At No. 2, the large bones exposed resemble those of *Atlantosaurus* with pneumatic cavities.[54] After supper Marsh told stories of his adventures in search of bones near Fort Bridger.[55]

June 5 Professor Marsh decided to stay another day with us. We started after breakfast for the lake. Professor M called attention to the little crustacea *(amphipods)* which live in the long waterweed which grows so abundantly at the bottom of the lake. We halted at the old quarry on the north shore [quarry 1½]. Marsh was much interested with the medium sized sacrum that lay exposed. He considered it to belong to the *Allosauridae*.[56] Marsh intended to have this quarry developed. From the lake we walked along the railroad track in a westerly direction to the bridge where the desperadoes had purposed wrecking the train.

We crossed many fine outcrops of red and yellow strata representing the Cretaceous and Jurassic group down into a hollow where the desperadoes encamped, a spot well adapted for their purposes. We sat down:

*Marsh Lunches with Reed and Ashley near Robber's Roost, June 5, 1879*. Watercolor by Arthur Lakes. Courtesy of the Peabody Museum of Natural History, Yale University.

Marsh, Reed, Ashley, and I under the identical cottonwood that had sheltered the bandits. Some mourning doves had built their nest in the branches overhead and kept flying to and fro with whistling wings. I made a rapid sketch of the party as they sat under the tree.[57]

Thence we wended our way through a number of small ravines to the foot of the saurian bluff on the south to a spot above the first *Ichthyosaurus* quarry where the eagle had built his nest in the overhanging cliff near the *Atlantosaurus* beds.[58]

At the base of this cliff we stopped to examine some coarse brown sandstone from which Reed had obtained the jaw of a little mammal *(Dryolestes)*.[59] We lunched under the shelter of a large fallen block of sandstone.

As we passed over the strata keeping a sharp lookout for bones, we lit on some vertebrae and other small bones of a saurian. This led to an eager search by the whole party, Professor Marsh full of excitement and enthusiasm leading the way. Presently a jaw was found with small crenated teeth like those of the *Ignanodon* or the modern iguana, an herbivorous saurian. By a little digging an ilium was unearthed of which at Marsh's request I made a rough sketch. After visiting one or two other places we returned to camp.[60]

June 6 Professor Marsh left on the train for the Eastern states. Reed, Ashley, and I went to the new discovery which is probably that of a *Laosaurus*. Three fir trees grew in an isolated clump close to the quarry so we named the quarry "Three Trees Quarry." The wind was blowing hard and with every blow of the pick we were blinded with dust. We succeeded in unearthing a femur and other bones.

June 7 Windy weather as yesterday. It blew so hard and the dust was so intolerable that we left work early. The men washed their clothes in camp. News reached us of a great mining excitement in North Park[61] and Reed determined to go there.

June 8 I spent [the day] in my tent reading and writing. Carlin brought in a wild-cat kitten *[Felis lynx canadensis]*. A pretty little creature with long tufts of hair growing out of its ears and with long, sharp claws and the prettiest wildest little kitten face. It was very tame but when shut up it screeched lustily like a screech owl. It quickly whipped the dog and bit Tom cat at the station who fled from it in terror and when offered a bit of meat seized it like a fury and hung on it till it was lifted off its feet.

June 9 Reed's birthday.[62] I made a little colored sketch of the lake and painted bluffs for him to stick up over his cabinet of fossils in Nebraska in honoring of his discoveries, work, and hardships. Reed is an example of the fact that all education is not confined to those who have gone through college but much is learnt by keen observation.

We walked over to our quarry which contains this curious little *Laosaurus* dinosaur, whose hind legs and whose appearance must have much resembled that of a kangaroo only with small lizard like head and a long lizard like tail with the toes armed with claws an inch long. Its bones are hollow like those of a bird. It was about ten feet long.

Our work was exceedingly unpleasant. A gale of wind was blowing, blowing dust in blinding clouds into our eyes as fast as we uncovered the material, and dry clay besides numerous provocations from paper, books, packing material being constantly blown away and chasing over the bluffs. Notwithstanding we kept at it tooth and nail and soon uncovered a lot of bones. The thigh and limb bones lying together as they were deposited when the animal lay down to die or was mired in the clay of a primeval marsh. The bones of the foot and the claws were all found in place in a mass and a number of little vertebrae about the size and shape of small reels of cotton.

But our best discovery was that of portions of a jaw with teeth in the sockets. The teeth are like those of the *Iguanodon* with sort of serrated edge adapted to feeding upon herbs and plants. They are very pretty and retain their original enamel and are about one half inch long. I sketched the bones as they lay, under great difficulty, the wind nearly blowing my [sketch]book out of my hands as well as blinding me with dust. Measurement and sketches being over, we left early for camp fairly blown away by the gale.

June 10 To Three Trees, found more bones and portion of a jaw with teeth in the sockets.

June 11 Adjourned to a spot three miles East of camp and again used the hand car and flew along the track to a quarry where we had found the remains of a small crocodile.[63] These we found accidentally whilst Reed and I were trying to get a cross section of the hill with a tapeline, R struck his pick into some clay and out leaped a pretty little femur about six inches long, then a claw and a sharp conical tooth not serrated like the Dino's but evidently a crocodile's as appeared also from the vertebrae and other remains besides a number of shields or scutes full of pits forming the crocodile armor.

About noon heavy clouds began to loom up over the south and quitting our work we hastened to take shelter under the castellated sandstones on the top of the bluff. Under them we stretched ourselves like modern saurians awaiting the coming storm; watching the instincts also of a few butterflies and moths that came in for shelter closing their wings and hanging themselves under a protruding leaf. Then the pause and stillness and hush of life which portends a storm of the first calibre and few distant rumblings. A few drops of rain and then, with a deafening crash followed by a blinding flash, the storm struck full on the back of the cliff behind us; then a heavy hailstorm, stones as large as bantams eggs. The sight was very fine with the sun shining brightly on the showers of white grass. The heavy clouds with repeated flashes moving off to the northward over the wild rocky outlines, the roar of the thunder and the rushy sound of the hail and the sense of deep solitude. Then in a few moments the storm was passed;

the sun shone bright and a few faint notes of joy broke out from the song of sparrows in the sagebrush and all nature seemed refreshed.

Reed shot an antelope whose horns met in an arch. He appeared at the quarry with the bloody hams as usual round his neck.

We picked up a few scutes or bits of the armor of crocodile and returned.

On the way home I killed a jack rabbit with a stone but did not bring it to camp as the men are prejudiced against eating it preferring the little cottontail rabbits.

June 12 A cold drizzle too cold for stationary work so we went prospecting along the cliffs finding fragments of bones lying loose but unable to trail them to their home also fine teeth both like the *Iguanodon* and *Megalosaurus*.[64] The former belong to a herbivore called *Morosaurus* and are an inch wide, black as ebony and almost two inches long. The others are serrated and belong to *Allosaurus*.[65]

We were fortunate enough at last to strike a spot [not quarry no. 6] where the bones of a great monster were sticking out of the bank and in place.

Whilst halting for noon we saw an animal about the size of a dog skulking up a ravine. R was soon after him and the crack of his rifle stretched out a coyote on the grass. R feels peculiar delight in slaughtering these in revenge for their nocturnal howls which have so often kept him awake at night. Once when snowed up he was driven to cooking and eating one of these carnivores. We were enthralled in the discovery of a number of tiny vertebrae whose centre were hollow. They may prove to belong to pterodactyls or some unknown creature.[66]

Same storm as the day before only we were drenched and returned home to find our camp in the same condition the tent leaking badly on our bedding. However a good fire soon set matters to rights. The bluffs was running with streams white as milk.

June 13 Went to Quarry 6 and cut a narrow trench along the face of the bluff about one hundred feet long in search of bones but found little except a few small hollow biconcave vertebrae. The centrum was hollow and filled with clay like a cotton reel.

The bones we have found at No. 6 are probably crocodilian: their vertebrae were amphicoelian [concave on both ends] and averaged about one inch diameter. The teeth were one half inch long sharp conical and striated and somewhat curved. The limb bones rarely exceeded six to eight inches in length. With the remains were found a tiny mammalian jaw about one inch long. The scapula was very perfect.[67]

Fine weather, ground wet and sloppy.[68] Continued work. The head of George Leech, an Englishman who keeps the snowshed about six miles from Como, appeared over the bank with his little terrier. The couple spend a lonely life of it in his exquisitely neat cabin. He is our vulture for as soon as report reaches him that R has killed an antelope and got the hindquarters, then Leech goes in on his trail and picks up the fore limbs. From him I heard some English news of the Turkish war etc. We have not had any news for a long time and a newspaper comes in with unusual interest at this lonely spot. Men are very sulky when the mail is empty and "the girls have gone back on them." Saw a wolverine in the distance and followed a fresh deer track home. Our regular thunderstorm occurred. Reed and Ashley put in the day casting bullets and loading ammunition. Packed up a lot of bones [from quarry no. 3]: a humerus and femur both over three feet long together with a lot of black ebony vertebrae of a carnivore known by the deep indentations along the sides as distinct from those of a herbivorous saurian.

Also portions of skull teeth and bones of *Morosaurus* a great herbivore some of its bones are like those of a crocodile but its vertebrae are drilled with large holes to hold air and lighten their weight called pneumatic cavities similar to those found in birds of flight and in the skull of an elephant. The dorsal or back vertebrae were of the ball and socket kind. The hollow part pointing towards the tail and hence called opistho coelous or hollow behind. Some idea of the size of the animal may be derived from the scapula which was four feet three inches long. A sharp pointed claw or rather the bone inside the claw was six to eight inches long with a great hoof on it for five toes or claws.

This herbivore when alive was forty feet long and walked on all fours and was probably sluggish in its movements and was no great thinker for his brain was smaller in proportion than that of any known vertebrate.[69]

June 14 Saturday we returned early and put in the afternoon packing up bones and dispatching boxes and in finishing sketches of bones.[70] In the evening I strolled down to the lake. The wild ducks were breeding in the rushes and occasionally as old drake would flash out with a motherly quack and splash the water with a sham broken wing and her diminutive brood after her. The eared grebes build in great numbers here but I could not discover their nests although their little heads were dotting the water in every direction and making the lake resound with their cries coyeet, coyeet, coyeeto varied by the quack of a mallard. Now and then the surface would be broken by the head and tail of a musk rat. There is something inexpressively wild about a solitary prairie lake in this region but the music of millions of mosquitoes made me beat a home retreat and join the boys at the station house who were trying to beat off homesickness by singing old southern songs till the mail arrived.

Reed and Ashley came home late from Medicine Bow in gorgeous new blue shirts, white broad brimmed hats, with cigars, and a blue shirt for me. So time in a place like this ebbs along with few striking events except such as our natural history affords.

June 15 Wrote letters home. Bathed in the lake. Ashley had a chase after an antelope fawn.

Whilst I was writing, a spermophile came into the tent and found a roll of cotton batting. He caught it up in his fore legs and tried to carry it off rolling with it over and over on his stomach in a most comical manner. Failing to carry it off he began stuffing his checks with it till they swelled out to a prodigious size as he rammed the cotton into them with his fore paws much as a birdstuffer does the skin of a bird. Finally he trotted off dragging the mass of cotton after him at times completely enveloping himself in it till he reached his hole.

Besides the amphipod crustacea, there are white worms, and tadpoles also live in the alkaline waters of the lake. The banks of the lake which rise about twelve feet above it show evidence of having been formed when the lake was much more extensive. They are of clay and resemble the loess formation [a fine grained silt or clay thought to be deposited by the wind]. The men bathed and rowed about the lake.

June 16 Reed left for Fort Laramie on his trip to North Park leaving Ashley and I to go on with the work. Went to No. 3 and after cutting down some ten feet of bank exposed a small elongated caudal vertebra ? and a pretty little chevron and some other vertebrae. Also a carnivorous tooth and a large flattened bone thirty inches long probably a scapula? as below it lay a coracoid about a foot wide.[71] A thunderstorm came on and we sheltered under the rocks.

June 17 At No. 3 again, we uncovered several procoelian dorsal vertebrae two of them consecutive, with them was a tangle of rotten processes one of the vertebra goes about five or six inches across center. Found a small tarsal? carpal bone? an herbivorous tooth with a long conical core running up through it.

Also a sort of twisted bone near a flat broad bone which may be another scapula. A perfect chevron and a good caudal vertebra.[72] Most of the bones were decayed. Mocking birds sang sweetly near the quarry, also the little gray shrike.[73]

Carlin saw a field mouse run up a wall with five little ones hanging to her teats.

June 18 At No. 3. More vertebrae. The scapula was 26 inches long.[74] Coracoid lying close to the broad end of it.

Packed up specimens and uncovered several dorsal and caudal vertebrae. In following up these vertebrae we tunnelled under the bank which gave way, nearly burying Ashley and me.

June 19 Found a vertebra in a concretion and another near it possibly a broken portion of the sacrum. Ashley followed down a long string of small consecutive caudal vertebrae from three inches to an inch in length evidently the remainder of the long tail.[75] There were thirteen in all till we came to the last one forming the tip end of the tail. It was conical at one end and was about one inch long by one half inch wide. Also found a large perfect chevron. Sketched the coracoid, as part of it was rotten.[76] Its extreme width was fourteen inches. Found a bone ten inches long, possibly portion of the foot or toes.[77] Ashley climbed to the top of the bluff and saw two elk within seventy yards but we had no rifle.

George Leech appeared on the dump telling us that a big rattlesnake as thick round as his arm had struck his dog. We went with him to look for it as he said he saw it coiled up on the edge of a badger's hole. We found the dog lying down with his hind foot swollen to three times its natural size. He howled when we pinched it. The dog seemed sick and lay down. Afterwards it limped away. We could not find the rattlesnake. We carried several bags of bones to the railroad track. Letter from home with father's photograph.[78]

June 20 Packed bones. Got out a fine humerus.[79]

June 21 Made boxes and packed specimens.

June 22 Bathed in the lake. Read Darwin's voyage of a naturalist.[80] Walked with Ashley over the hogbacks to the south. The strata are in waves or upturned ridges representing the Dakotah No. 1, Benton, and Fort Pierre groups No. 2 and 3 Cretaceous. Deposits of iron ore occur in large nodules in the gray clays and shales of the Benton group.

June 23 The record for this week is one of prospecting and exploring and examining our surroundings.[81] Reed went to North Park and Prof Marsh wrote to me to send him a map and pictures of the country showing the position of our boneyards.[82] So I took Ashley with me and started on a voyage of discovery and survey.

We went westward following the outer rim of the remarkable fold which forms the leading feature of this locality. By some internal force probably by contraction of the earth's crust in mountain building the strata has been pricked up into a fold, thus

The top of the arch being scooped out by water and filled by the lake leaving the broken ends on either side telling of the natural arch that used to be. It was part of the object of our ramble to trace out this fold. So we walked along the railroad track for a long way by the side of a perpendicular wall with the top broken off and jagged. After a while when we reached the bridge where the robber's tried to wreck the train the strata curved round in fold upon fold like the outer ring of an amphitheater.

At this point was the trestle bridge where the desperadoes proposed to wreck the train in a ravine about twenty feet wide by thirty deep. The banks on either side cut perpendicular by an old water course. The ravine is spanned by a little railroad bridge. The object of the bandits was to pull out one of the rails to which they had attached a bit of telegraph wire. Plunge the train into the ravine and then plunder the cars amongst the dead and dying. For there is no saying how many would have been killed by this devilish scheme. They could not have chosen a better spot for their work or concealment. The folding of the strata forms many little ravines and can be completely shut out from view and in one of these in a most picturesque spot the banditti encamped under some trees known as Robbers Roost. Marsh and our party eat [ate] lunch under the same trees.

Ashley who was ahead of me startled a fine buck black tailed deer out of a ravine. Like a contrary emblem to the deeds of violence, a couple of mourning doves had built their nests in the crotch of the cottonwood tree and kept up a perpetual mourning and cooing all the time we were there.

From this point we had a fine view of the country west of us. A broken rolling country desolate and wild looking in the extreme with no farms or tilled lands to cheer the eye with the thought of homes and civilization.

In the distance an alkali swamp was blowing dust like a cloud of smoke. The little village of Medicine Bow could be distinguished along the railroad track and the distance was shut in by low mountains just reaching timberline haunts of bands of elk, deer, bear and deer. One of these hills is called Freezeout Mountain where a party of hunters was once snowed in. Another is Elk Mountain.

On climbing the ridge over our quarries we found a pile of stones laid there by Clarence King's survey[83] and near it some quartzite Indian arrowheads. The cliffs above the lake are green, lilac, purple and red sculptured with ravines. Below as was a layer of sandstone where the jaw of a small mammal *Dryolestes priscus* a species of marsupial, like the kangaroos of Australia, was discovered by Reed. This was a great prize as it is one of the very first that has been discovered in the Upper Jurassic period. Only

*Como Bluff.* Reed and other diggers are looking south at the main bluff. Quarries no. 9 and 10 are to the far left. Watercolor by Arthur Lakes. Courtesy of the Peabody Museum of Natural History, Yale University.

saurians and similar monsters were supposed to have lived in that age. This was a little fellow probably an insectivorous animal not much larger than the kangaroo rat.

We endeavored to make a section of the cliff occupied by the golden eagle's nest. The descent was slippery and wind blowing a gale which it does two days out of three and at the first measurement snapped the tape line. However we slid down the cliff over the gray clays where the big saurians lie entombed, down over the variegated beds above which the eagle's nest is perched and finally reached the bottom without difficulty. The sculpturing of these purple cliffs into bench upon bench reminded me of pictures we see of the bad lands of Dakotah. This is due to water.

Passing on we went to our little *Laosaurus* quarry, the little bird like saurian. Thence to the tomb of the toothless *Ichthyosaurus, Sauranodon,* or *Baptanodon*. The *Ichthyosaurus* so plentiful in Europe has never been found before in America but this creature unlike its relatives had no teeth. The remains were found amongst quantities of belemnites which look like petrified cigars scattered about in great profusion. All these remains are separated from one another by different hills of strata like the stories in a house. In the top story was the remains of a primeval forest of trees like our own. Below lie the *Morosaurus* beds. In No. 2 and below them those of the little *laosaur* and crocodiles and below that the *Sauranodon*. Each set of strata is divided from the other by a belt of sandstone.

We were glad after our days work to shake off our fatigue by a bathe in the lake which was swarming with living saurians viz axoloti or water newts. They would squirm out from under our feet at every step in the mud and we seemed swimming through lizards.

June 24 The boarding house boss's wife being sick of mountain fever,[84] I had to leave Ashley at home to cook for them and taking a young boy[85] with me was driven on the hand car to the east end of the basin to a spot where snow fences are built to shelter the track in winter from the tremendous snowdrifts. Even with them the line is blocked often and then comes the great steam plow with its beak like the bow of [a] steam ram plunging into the drift with two or three engines behind it to back it and make way for liberty. The fences are simply boards nailed together like a fence. Leaving the car we walked down into a deep ravine formed by Rock Creek which wound around through pleasant meadow land like a serpent. On its opposite bank were fiery red Trias rocks lying tier upon tier and stretching away till the whole plain seemed like a plain of fiery red rock. On the other bank were the castellated cliffs of the Dakotah.

Whilst I was examining the strata, George startled a couple of elk which sprang across the ravine with branching horns carried well in front. They swam the river and their hoofs might have been heard clattering over the sandstones of the plain of red rocks.

Under the eaves of these red rocks where it had been undermined by the river a colony of cliff swallows had built their clay nests like a honey comb packed close together, whilst clouds of the anxious proprietors filled the air with their wings and cries.

We reached Quarry No. 4. A great trench of eighty feet long dug in the side of a domelike bluff of gray and purple clay by Reed and party last winter.[86] The bone diggers evidently lived on elk to judge from the remains lying round mingled with the prehistoric bones.

The clays are here cut by deep narrow ravines filled with luxuriant grass, the lairs and feeding places of elk. We dined at the snow fence and afterwards explored the head of the fold. On a summit of a rocky eminence we found a rude Indian fort made by stones placed in a circle in the center of which a larger sage bush was growing showing how old the fort was.[87] I found some Indian axes and arrow heads.

After dinner we climbed the bluff which curls out from the Dakotah and commands the upper end of the Como basin. Near the base of the bluff we encountered the *Sauranodon* beds, dipping east between them and the Dakotah ridge are the variegated and *Atlantosaurus* beds much worn away by erosion.

The bluff is capped with the heavy sandstones of the Dakotah on which is a wide Indian rifle pit or fort made of a circle of loose stones. It is about twenty feet wide. From this point we have a fine view of the Como basin. A coyote skulked out from under the rocks. The Dakotah dips E. at about an angle of 20°.

June 25 I sent George to the Quarry [no. 3] to dig and went alone to examine and follow the strata along the North shore of the lake.

I measured a section across the uplifted strata forming the north bank.

The first bed from the lake is the *Sauranodon*, followed in the north by the usual succession. At the edge of the lake is some course red sandstone. In this I thought I noticed obscure signs of vegetation which is scarce in the Red Beds.

Walking along the crest of the Dakotah I looked down into the valley of Rock Creek.

The thickness of the Dakotahs is about two hundred feet. The lower section of it is deeply impregnated with iron oxide. Looking west I could follow two ridges of the red Trias forming a continuation of the lake shore strata also two more southern outcrops of it. The dip is south but the inclination is at a more acute angle.

A few hundred feet further brings us to another ridge where the angle is yet smaller and not much further on we should reach the dip of the arch and find it nearly flat.

But continuing along this same ridge of a dark red and grayish shale I came to a point not far from where Rock Creek takes a bend and cuts through the Dakotah forming an admirably exposed section from Trias to Dakotah. In one of the ravines two antelope jumped up almost beneath my feet and bounded away like the wind. Being very hot, tired and thirsty, I was glad to see the cool stream of Rock Creek winding through the strata near the point where they began to curve round as at Robbers Roost. I turned down into a little ravine and found a crystal brook clear and cold issuing in a small waterfall from a grassy bank. The ground around it was much trampled by deer tracks showing it to be a favorite drinking place. On the bank I found a small stone spear head of Indian workmanship.

On reaching Rock Creek I had a fine view of the cross section of the strata from the almost vertical Dakotah to the nearly horizontal Trias. The contrasts of color are very fine. The gray Dakotah, the purple and variegated Jurassic, the saffron yellow sandstone capping the sauranodon beds, the massive cream colored sandstone of the Jura Trias and next to it the deep red Trias sandstone variegated with streaks of darker red, gray and white. The latter is from belts of gypsum.

I startled a wild duck from its nest in the rushes and also a large horned owl from the bank. On the west bank of the river along the edge of a cliff is another outflow of springs issuing from the Trias rocks. Near there I found many flowers of the *Dodecathron* [*pulchellum* (western shootingstar)]. I climbed to the top of the table land of horizontal Trias above the springs. The red rock is capped with a bed of gypsum and mottled shale nearly horizontal.

From the top of this table land I looked down into a little park-like basin caused by the dish-like curvature of the end of the fold and the nearly horizontal beds of the Trias forming the axis or summit or the arch.

About the middle of this park was a little cabin built over some remarkable springs which bubble up in several little ponds with a continuous ebullition of sand. The largest is in the center of a large pond about three feet deep. It boils up with great force, the column of waters over a foot wide, and brings up with a continuous cloud of black substance resembling peaty mud and also leaves an encrustation like calcareous tufa [a calcium rock deposit by springs]. A water snake was coiling in and out of this black cloud and quantities of suckers were lying on the bottom of the pond watching the troubling of the waters. Carlin built a shanty here with a view to raising trout.

On crossing the park at the south end we again meet the Trias rocks dipping on a contrary direction toward the south showing that we have passed over the top of the anticlinal arch.

Rock Creek passes down through the park and enters the valley where No.4 quarry is situated.

Here at the SE end of the ridge we found the *Sauranodon* beds dipping at a slight angle and further on the Dakotah.

Leaving the valley of Rock Creek the strata dip gently from the centre of the Como basin through the center of which runs the railroad forming a grassy slope on which several herds of antelope were grazing. Tired and hungry I was glad to arrive at No. 3 and found George with dinner ready for me.

June 26 In the evening Reed's blankets arrived and about midnight Reed himself came in.

His trip to North Park had proved unsuccessful.

He had found some saurian bones at Laramie which were being excavated by young Wanness a former pupil of mine. Reed recognized a sketch of Jarvis Hall in Wanness's house I had made for Wanless some nine years ago.[88] Reed had seen herds of antelope accompanied by their fawns and could have caught as many as he liked of the latter. When they arrived at the Park they found there was a panic on account of the Indians.

Reed camped with a man and went out antelope hunting whilst shooting as many as he wanted he noticed an Indian perched on every high point.

They came into camp and said "heap hungry." Reed gave them some food. They said, "Too much white man in Park, etc." A woman being much scared added to the panic and there was a general stampede till the park was nearly abandoned except by some old frontiersmen.[89]

June 27 Reed being too tired, rested in camp whilst Ashley and I went over to the bluffs east of the station and spent the day prospecting among the rounded clay bluffs to which in places we had to cling like beetles. We dug out a good many waterworn fragments of bones and several crocodile's teeth some very small, others an inch long. We also found some carnivorous dinosaur teeth and a few herbivorous ones. There were no signs of a skeleton. It happened as if a medley of fragments had been washed together by running water in connection with some concretions.

June 28 Reed and Ashley opened up an extension of Quarry No. 3 but without success.

I left them to make a sketch from the Indian fort. Several curious mushroom monuments are made by erosion of the sandstones of the Dakotah which near the fort, being the curling round of the fold, are nearly horizontal. I found near the fort two fine quartzite arrowheads. Doubtless the Indians made arrowheads here as there were numerous chips of quartzite lying about together with the rude stone implement with which they are supposed to have cut them out.

The rude fort or rifle pit was made by building a low curtain wall of loose stones. It was about twelve feet diameter and about four feet high. In its center grew a very large sage bush which shows the antiquity of the structure as the sage bush grows with infinite slowness. As Reed remarked it is a curious fact that no living man ever saw a young sagebush.[90]

A nighthawk *(Chordeiles) [minor]* started up from my feet and whilst sketching kept swooping close by me; probably her eggs or young were near. These birds will be close among the barren rocks trusting to their color so much like that of the rocks to conceal them. Their eggs are also hard to find for the same reason, their color resembling granite. They will sometimes alight close to you and allow themselves to be pelted with stones before they will rise or fly off. They fill the air with their harsh screeches and with the peculiar sound caused by their wings suddenly filling and rising again.

June 29 Read Darwin. Men brought in six young teal just fledged.

June 30 Stayed in camp to make a sketch in pen and ink of the bluffs of Como from the east shore of the lake. The sketch was three feet long and intended for Professor Marsh.[91] It was a fine morning and I sat under the shelter of the railroad station.

A spermophile was brought in alive for the benefit of the wildcat kitten who pounced upon it, though it was as large as herself, with the greatest ferocity. The spermophile caught pussy by the throat with his sharp rodent teeth but the little cat never let go her grip till she had killed it and then waddled off with her prey dangling between her legs with an air as much as to say "This is my funeral" her ears erect and her whole body vibrating with ferocity.

Reed and Ashley returned from No. 8 reporting the discovery of an enormous vertebra, also bringing in a huge sabre-like carnivorous tooth which I drew.[92]

July 1 Went to No. 8 and saw the bone. It appeared to be a very long vertebra, sixteen inches long with a keel down the center of the centrium and large arches formed by transverse processes. There were the remains apparently of another vertebra attached to the hollow end of the preceding one. The men also found about five feet below it a heterogeneous mass of bones: little toe bones, big carnivorous teeth, small elongated caudal vertebra, and a good sized carnivorous vertebra (a caudal) [these were from an allosaurus], also a carpal or tarsal of some large dinosaur [probably a sauropod], also six or seven inches of a good sized jaw with many small herbivorous teeth slightly crenated on the edge and with long fangs,[93] some crocodiles' teeth, and waterworn fragments. Also a vertebra of some large saurian a foot in diameter. The stratum was full of little round concretions. It seemed as if the deposit had been made by running water.

July 2 Stayed in camp and drew picture of Como basin four and half feet long.[94] Men returned reporting new discoveries east of the sandstone bone stratum: small crenated teeth, crocodiles' teeth, etc.[95]

July 3 Men remained in camp and built a boat in half a day. We christened it the *W.H. Reed*. Carried it to the lake and launched it.

I made a sketch of the men in front of the tent and also put in figures of Reed, Ashley, and myself in the foreground of my picture from the Indian fort and sent the picture to Marsh.

Reed came back from the lake with twelve dozen eggs they had collected from the grebes nests on the rushes in about an hour. In former years as many as seventy five dozen have been collected in a day.

July 4 Our first announcement of the day was the boy[96] coming into our tent at 6 o'clock to awaken us. Reed sat up in bed and shot a hole through the tent with his revolver. Then Ashley went out and fired off his six shooter.

After breakfast we all started out to shoot rabbits with our pistols. Numbers of bush and jack rabbits were startled from their cover in the sage brush. The jack rabbits would run a few yards and then stand up on their hind legs to look back and offer a fatal mark for the revolver. At other times they tried concealment by squatting down and throwing their long ears on their backs. Some six or seven rabbits were soon bagged.

The weather was lovely and still. I crossed over a little rise separating Como from Rock Springs, i.e. the spring which feeds the railroad tank. I put up two jack rabbits an old and a young one and sat still watching their movements. They would run a few paces, then standup on their hind legs vibrating nervously their fore paws with ears erect listening. After a while as I remained motionless they continued their feeding.

To keep perfectly still and quiet often assures wild animals and allows one to observe their domestic habits.

Walking onto the edge of the lake, I peeped over the bank. The water was clear and smooth as glass. The red and gray rocks on the banks reflected in it. Thousands of siredon lizards [tiger salamanders] were swimming lazily near the surface or probing the water weed near the bottom in search of food or lying still and basking in the warm shallow water near shore. I startled a wild duck beneath the bank who swam out with her whole brood of little ducklings leaving a long ripple behind them. The old bird encouraged the young ones by a low purring sound.

Several cow black birds *(Molothrus picoris)*[97] with their sleek black satin plumage were stalking statelily along the edge of the water reminding [one] of our English rooks. I sat perfectly still on the bank. Presently two tiny striped chipmunks appeared from their holes closely and after much nervous flirting of their tails the little elfs came right up to me to gaze with admiration upon a pair of very gawdy carpet slippers that I wore. They seemed less afraid of me than of the movements of a rabbit that was ambling along the bank.

I noticed how much sympathy so to speak there is amongst animals and how like their ways are to those of men. The stately blackbird as will in his way meet a more humble cowbird on his walk along a narrow strip of shore; he declined to yield the path to the humble bird who had to fly out his lordship's path.

The distant report of a pistol caused every inhabitant of the lake to prick up its ears. The little chipmunk mounted his sentinel rock to see what was up. The blackbird paused in his stately constitutional promenade and uttered a note of warning. The ducks quacked and a general sense of alarm or "qui vive"[98] pervaded the hitherto peaceful lake. Soon all was quiet again and peaceful. A few copper butterflies and an argus-eyed meadow brown fluttered about among the reeds near the shore.[99]

On returning to camp I found the men had been doing justice to a beer keg freely tapped for all hands beneath the water tank. George Leech who had arrived from his snow shed, assisting in the operation. At noon we were all invited to the stationhouse where Mrs. Chase, the stationmaster's wife, had prepared a Fourth of July dinner for us which we all enjoyed. It seemed quite homelike, custard and other pies and canned fruits were the luxuries of the table. After dinner the men again adjourned to the tank and again sat down affectionately around the beer keg. I read extracts of Pickwick to the men in the tent.[100] Made a sketch of the party and tent. I went to bathe whilst the men slept off the affects of the beer keg.

In the evening Mrs. Chase was unwell. After sunset a chaise and pair of horses arrived and deposited a midwife at the station and at 9:05 PM Mrs. C presented Como with a Fourth of July baby daughter.[101] We had rabbit stew for supper and so ended the glorious 4th and the moon came up a red ball over the bluff.

July 5 Went to Quarry No. 8 about 400 yards east about 200 yards from Castle Rock to the low bluff from which the upper surface had been removed by erosion near the base.

We found turtle or crocodile bones, crocodile scutes, striated crocodile teeth one quarter to one half inch long, small vertebra one inch long flat at both ends, others slightly concave at both ends about one inch long, a curved limb bone four and half inches long—bone like a straw striated, an herbivorous tooth, a small fish vertebra.[102]

The stratum is soft-gray shale underlaid by concretions, the slope very, very gentle, a large surface could easily be uncovered.

July 6 Read Darwin. Men went out hunting and prospecting and brought in some marine cretaceous fossils from the Fort Pierre group among them a shark's tooth like *Ptychodus montoni* and some lancet-shaped selactian teeth serrated [of the order *Selachii,* which includes sharks and rays].

July 7 Went to No. 9.[103]

*Quarry No. 9.* The most important Mesozoic mammal fossil discoveries of the nineteenth century were made here. The natural monuments at the upper right were called the Indian Fort. Watercolor by Arthur Lakes. Courtesy of the Peabody Museum of Natural History, Yale University.

July 8 Same as on the 7th.

July 9 Reed crossed the lake to Quarry 1½. Ashley and I to No. 9, found several sabre shaped serrated teeth; also a crenated ignanodon tooth.[104]

July 10 At No. 9 found small jaw half inch long, probably mammalian, teeth indistinct, also part of another jaw. Reed found remains of a *Sauranodon* near Robbers Roost.[105]

July 11 We went with Reed to excavate his *Sauranodon*. We paddled across the lake whose waters were clear and smooth as glass. The bottom was thickly covered with waterweed some of it like sea grass others but little tufts like horsetail rushes. Saw a few siredons swimming amongst the weed.

About fifty yards from shore as we approached there was a grand splashing of grebes scudding across the surface; as we drew nearer, we found that alarmed by the recent raid on their nests among the reeds near shore, they had actually built a colony of floating nests out in the middle of the lake anchoring them to the weed at the bottom of strands of grass or tying them to the beds of the horsetail rushes which just peeped above the surface. There were about a dozen nests made of the fine hairlike water grass. This little cushion floating up and down on the ripples bore its cargo of two or three snow white eggs about the size of those of a pigeon. It was a strange and pretty sight.

With our picks and shovels over our backs we soon reached the Robbers Roost at a point about a half a mile west of Como station. We began operations finding a few calcareous concretions characteristic of the *Sauranodon* beds and also a few disc like vertebrae lying on the surface. We soon dug out the skull of a *Sauranodon* tolerably perfect. It was about two feet long, its long snout penetrating into the ground and detached from the skull. After dinner we renewed our search and opened up a trench.

We were rewarded by getting out a large number of these concretions which contained several bones and also another perfect skull four feet long showing the large saucer-like eye and two strings of consecutive vertebrae all more or less united by concretionary matter; also a large limb bone (scapula?). We succeeded in getting out about eight feet of the animal's length. The men were in high spirits over the discovery. I made a sketch of the skeleton as it lay. There appeared to be a pair of animals in the same hole as we found two skulls.[106]

July 13 I spent the morning near the lake reading and watching the habits of the grebes.

I crept up to the edge of the bank and lay at full length in the grass where they could not see me.

There were now some sixty or one hundred nests floating on the water about fifty yards from shore and on many of them the parent birds were sitting whilst others were swimming close to the shore repeatingly diving. On my showing myself, there was a general splashing as they dashed from their nests and sped over the water or rather through the water splashing it with their short wings. It reminded me of a discharge of grapeshot ricocheting over the water. After a time one or two single birds would return, swim near to their nests and around them, halt for a moment as if to steady themselves for a spring, and then a sudden jerk hop up on the floating nest. Then they began to arrange the nest or remove some parts of it with their bills and then settled down on their eggs. Others taking courage kept coming in till each little boat was manned with a duck its head and neck just appearing above the nest. Their coyeet! coyeet! cries ringing over the lake. Saw flocks of *Phalaropes* wading up to their bellies in the water and a tall black and white avocet among them.[107] Tried to wade over to the grebes' nests but the weed was too thick.

July 14[108] We all went out prospecting for Sauranodon bones in the direction of Robbers Roost keeping close to and following out the *Sauranodon* bed horizon. We found many calcareous concretions full of small bivalve shells.

At the lower portion of the bed is a shaley calcareous sandstone full of shells especially of *Ostrea marshii, Ammonites cordatus* and apparently the spires of an echinus *(hemicidaus?)*. The sandstone capping the *Sauranodon* bed is most beautifully ripple marked and breaks in their slabs. A deep gulch separates the *Sauranodon* beds from the *Atlantosaurus* horizon.

At Robbers Roost is a magnificent display of strata. Many small faults are produced by the tension of the strata in curling round. Crossing a deep ravine we found fine specimens of *Belemnites densus* at the base of the *Sauranodon* beds close to a coarse brown limestone full of large crystals of iron pyrites.

We followed round the curve of the fold and return home along the *Atlantosaurus* horizon. The beds intermediate between the *Sauranodon* beds and the cream colored sandstone of the Trias consist first of a gritty coarse gray sandstone. Below this some red shales and then the cream colored sandstone.

Reed found a perfect phragmocone of a belemnite.[109] We ran our *Sauranodon* bones on the hand car to the station and packed them off to Marsh.[110]

July 15 Having packed off *Sauranodons* Nos. 3 and 4 and prospected the hogbacks NW of Como towards Robbers Roost, we determined to thoroughly explore the *Sauranodon* beds across the lake towards the NE. We started early on Tuesday morning, Reed, Ashley, Reed's little boy, and myself, the little boy carrying my revolver. On our way we found some rude Indian implements some stone axes and some polished quartz pebbles probably used by squaws for dressing deer skins.

We reached Rock Creek gap where Rock Creek cuts through the Dakotah and Jurassic ridges. We left our dinner at a spring and waded across the river to the *Sauranodon* Jurassic beds. At the base of these we found several *Ammonites cordatus* and other marine shells in concise sandy shales.

I left the party and climbed to the top of the Dakotah hog back: Two golden eagles rose near me as I approached. In the upturned sandstone of the Dakotah ridge I noticed a peculiar formation. It resembled a very thick water wheel black with oxide of iron and balanced on the top of the ridge, a prominent object. It proved to be an enormous circular concretion of sandstone deeply impregnated with iron oxide. It was five feet wide and four feet thick, left in this prominent position by the erosion of the surrounding sandstone on a loose block of sandstone.

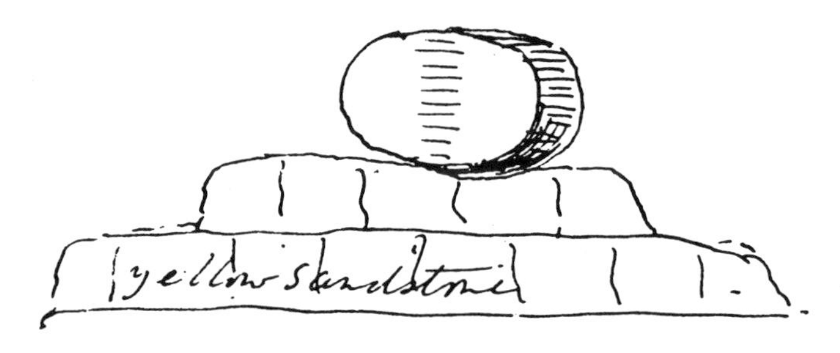


I saw traces of a plant fossil like phragnites jointed.

Recrossing the river I sat down to sketch the fine display of upturned strata on the opposite bank of the stream.

These show many varieties of color and texture:

1. First, the dark gray shales of the Fort Benton group [Upper Cretaceous].
2. The yellow Dakotah sandstone [Lower Cretaceous].
3. The variegated gray greenish and purple *Atlantosaurus* beds of the Jurassic [Morrison Formation, Upper Jurassic].
4. The white and brown sandstones intervening between the Atlantosaurus and *Sauranodon* beds rising in vertical walls of masonry with a deep ravine between them and the *Atlantosaurus* beds [Jurassic].
5. The sulphur yellow sandstone capping the *Sauranodon* beds and a gentle slope of green turf covering the gray calcareous shales in which the *Sauronodons* lie embedded [Middle Jurassic].
6. Heavy masses of massive cream colored crossbedded sandstone forming the base of the Jurassic and capping the red Trias [Lower Jurassic].
7. The red Trias conglomerates with some gray sandstones between

them. These gradually decrease in dip till they assume horizontality where they form the top of the arch of the fold. This is shown by a gentle slopping ravine and then by almost flat topped table lands. The cap of one of these table land is of gypsum resting on the red sandstone.
8. Then a soft pack basined in by red Trias rocks.

At noon the men came back from their prospecting to dine by the side of the brook. I sketched them at dinner. They brought with them as trophies about a dozen curious little bones somewhat like small flat vertebrae and not unlike some cyathrophylloid Silurian corals [family of fossil stony cup corals (order *Rugosa*)]. We guessed at their being little bones of the *Sauranodon*'s paddle. They averaged from an inch to half an inch in diameter both top and bottom were roughened like the cap of a Silurian coral. The sides had two parallel rows of little tubercles like bony warts, the center ones being the larger. All the surface of the bones was of a rough concretious character.

After dinner we decided to again visit the bone locality. We passed our way through the park and visited the bubbling springs. We roused a flock of gadwall ducks from a pond and an old bird with half a dozen ducklings all of which seemed very tame.

We crossed the river by means of a beaver dam and found ourselves on a plateau of horizontal red Triassic sandstones interstratified for from thirty to fifty feet with beds of gypsum, the white of the gypsum in striking contrast to the red layers of sandstone.

Hearing the cry of a bird I followed the sound and came upon a mountain plover. It was feigning the broken wing dodge but [I] failed to find either its nest or its young. After digging for sometime where the men found the paddle bones, we failed to find any more.

I made the following section of the *Sauranodon* beds at this point:

Top
L. Jurassic *Sauranodon* beds 100 feet
1. Gray limestone
2. Yellow sandstone
3. Gray shale
4. Covered space
5. Cream colored and white very soft sandstone [crossed through]
Triassic
6. White and cream colored very soft massive sandstone

7. A streak of violet shale
   Variegated gray and red and yellow sandstone
   Red sandstone capped with fifty feet of gypsum

We returned by way of the ravine occupied by the *Atlantosaurus* beds. (1) The yellow sandstone of the Dakotah (2) has as its base a bed of fine hard gray conglomerate composed of small hard black chalcedonic pebbles [a translucent grayish quartz] (3) beneath which is a bed of fine dark slate colored shale looking almost black and probably contains lignite. (4) Below this is the coarse brown sandstone noted for containing a mammalian jaw at Como. It is separated by about twenty feet from the Dakotah conglomerate.

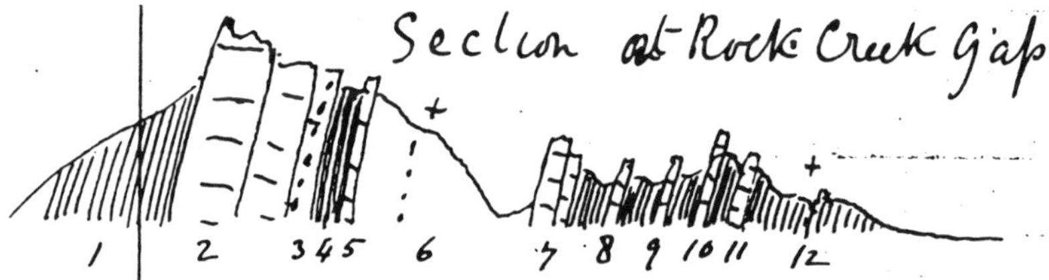

Section at Rock Creek Gap

1 2 345 6 7 8 9 10 11 12

| | | | ft |
|---|---|---|---|
| Cretaceous | 1 | = Fort Benton shale | 300? thick |
| | 2 | Dakotah sandstone | 200 |
| | 3 | Conglomerate | 20 |
| | 4 | Blackshale | 20 |
| Upper Juras | 5 | Mammalian Jaw Sandstone | 5 |
| | 6 | Atlantosaurus beds | |
| | 7 | Soft White sandstone | 20 |
| L. Jurassic | 8 | Limestone | 5 |
| | 9 | White Sandstone | 5 |
| | 10 | Sulphur yellow Sandstone | 5 |
| | 12 | Sauranodon Shales | 100 feet |
| | | Sandy Shales & Ammonites etc. | |

We returned by way of the springs which supply the railroad tank passing on our way an alkaline deposit as white as snow. It is a dried up lake. The men set fire to the rank grass growing beside it. The water of the railroad tank is very good.

July 16[111] Resumed work at No. 9 finding some small jaws averaging one inch to a half an inch long with teeth [in] them resembling crocodiles. [These were mammalian.]

July 17 Found some loose vertebra like *Laosaurus*.

July 18 Prospecting.

July 19 Prospected for *Laosaurus* bones along the bluffs S.W. of station from Three Trees [Quarry] west. Found a few caudal vertebra and a solitary *Sauranodon* vertebra and some turtle bones.

July 20 Read and wrote letters and bathed.

July 21 Worked at No. 9. Reed found an enormous carnivorous dinosaur's tooth six inches long and two inches broad serrated on both sides. Also some smaller ones like it.[112]

July 22 Sprained my back, returned home. More jaws found.

July 23 Stayed in camp. Men found more jaws.[113]

July 24 In camp.

July 25 Made colored drawing of section at Rock Creek gap.[114]

July 26 Men came back with report of discovery of very big bones at a spot between Quarries 8 and 9.[115] Heavy thunderstorms hailstones fell the size of hens eggs. Telegraph wires broken. Made sketches for Reed and Kennedy.

July 27 Read Darwin, wrote letters. Men cut hair. Siredons came up out of the lake and waddled into camp.

ey and I worked at the new discovery, No. 10. We found closely packed together. We got out part of what appears of enormous size but split through the middle. One of the ifteen inches in diameter. We also found some large limb bones, a large chevron, and some caudal vertebrae. Evidently a huge monster in there.

July 29 Got out the sacrum at No. 10.

July 30 Tried to arrange the scattered fragments at No. 10.

July 31 Worked at [no.] 9 and 10.

August 1 At No. 9 found a jaw two or three inches long with teeth, probably mammalian.

## Yellow Shanks; Curiosity of Birds; Cope's Arrival

I went down to the lake to bathe. Numbers of long legged "yellow shanks"[116] and smaller sandpipers were wading up to their breasts in the water.

When I began to bathe their curiosity was excited and they came up so close to me I could have touched them with an oar. They stood balanced on their long legs, their forms and shadows reflected in the smooth water bowing and nodding to me in a comically polite manner.

They were evidently attached by curiosity and showed no sense of fear of a naked man.

I stood perfectly still watching them quietly feeding close to me. One of them calmly laid his long neck on his shoulders and tried to take a nap. A rude or mischievous smaller sandpiper stepped up and gave him a shove like boys do to one another when they fall asleep in church.

The yellow shanks awoke with a scream. These birds which have only arrived within the past week were wonderfully tame or unsophisticated. Flocks of wild geese and other wild fowl are beginning to come into the lake.

On returning from my bathe, the train arrived and a tall, rather interesting looking young man stepped out of the coach and introduced himself as Professor Cope.[117] He brought his blankets and a rubber bed for camping excursions.

Edward Drinker Cope.
Courtesy of the Smithsonian Institution Archives.

August 2 With Cope at breakfast and had a pleasant chat with him about England. He entertained his party by singing comic songs with a refrain at the end like the howl of a coyote. I went across the lake to wrap up bones at No. 1½. Was caught in a heavy thunderstorm and drenched.

After supper chatted with Professor Cope about geological matters. He considers the Florissant Basin (South Park) as either Lower Miocene or Upper Eocene.[118] The fishes found there are of three or four kinds: some of them are *Amias* (suckers); the larger ones are *Ganoids* allied to those now living.

He has had a large restoration made of the *Atlantosaurus* in the grounds of the Philadelphia exposition. Three acres of the grounds were allocated to him.[119] He left shortly after by train to San Francisco.

August 3 In camp as usual.

August 4 Rain.

August 5 Found a mammal jaw at No. 9 one half inch long with crenated iganodon teeth![120]

Wild geese baying on the lake like a pack of hounds.

August 6 We went to Robbers Roost and found signs of bones about five hundred yards west of the railroad bridge. One large bone was two feet long and broad at one end measuring fifteen inches with a deep hollow in it shaped like a spoon and was a limb bone probably a humerus. Also a femur four feet long six inches long. It was long and slender. Sketched it before it broke up. We only saved the distal end. Close to it was this flat bone like a coracoid (perhaps a plate of armor) [it was a dermal plate]. This also broke to pieces. We found two or three medium sized vertebrae. Some of the smaller bones seemed to be hollow.[121]

We examined the beds at the base of the *Sauranodon* bed and found a fossil like a coil of small worms resembling *scapularia*, also a fragment of an echines *(Hermicidares),* fragments of pinna, and apparently some corals. Also, *Ostrea marshii* [a type of oyster] and some bivalves striated like a cardium; with these were worm-like fragments possibly crinoids [sea lilies].

Reed killed two deer and two antelopes. Saw great numbers of horned toads. The young ones were about two inches long.

August 7 Went out prospecting with Ashley. Reed went on horseback to bring home his game and was thrown and kicked by his horse. A and I found signs of a *Sauranodon* near Three Trees [Quarry], a vertebra four inches diameter in a concretion.

August 8 Reed laid up. A and I went to Robbers Roost to Quarry 12 and dug out a tibia and fibula united. We spent the afternoon cutting down the bank and enlarging the excavation. Ashley missed his little pick he had left at the quarry.

August 9 Reed came with us to No. 12. We got out several bones. The tibia and fibula lay close to the femur. So we have thus far got out at least one leg except the foot. We uncovered part of a broad flat bone. Sent drawings to Marsh.[122]

Ashley again missed his other pick which weighed about four pounds. After hunting about we found one pick twenty yards from the quarry and the other at about the same distance near the top of a little hillock. Teeth marks on the handles showed that the robbers were the mountain rats.

Packing boxes to Yale.[123]

**August 10** Bathed in the lake in company of the sandpipers who played around me about ten feet off. They paid no attention to me but the shadow of a marsh buzzard[124] caused a scream of alarm and a general hurried flight. Geese and ducks are plentiful and the men are out shooting.

**August 11** My tent arrived and was pitched near Reed's.

**August 12** Worked at No. 8. Found a tibia quite perfect and hollow eleven inches long.[125]

**August 13** At [no.] 8 found a toe bone and a vertebra.

Went to No.11 and sketched big humerus (forty-five inches) with it there are eight vertebrae, one measured eleven inches diameter. Some are flattened and not very thick, others are caudals. Got out a phalanx bone ten inches long.[126]

In the evening after a shower of rain, camp was besieged by siredons. They came up like the Egyptian plague of frogs[127] in thousands from the lake crawling in every direction and over the floor of my tent insinuating themselves under my bed and remonstrating vigorously with their tails when caught. I threw out twelve before I went to sleep. And still they came waddling in. What with lizards squirming under my bed, mice running over my blankets and gnawing at my books and papers, and wild geese screaming from the lake, the visitors and "Voices of the Night" were a little too much for me.

**August 14** Went to Quarry 11. Uncovered a large mass of bone. We prospected the *Laosaurus* horizon on our way home finding a *Laosaurus*? vertebra. Also indications at two localities of very large bones on the *Laosaurus* horizon which lies several feet below the ordinary *Atlantosaurus* horizon. We found some turtle bones in the *Sauranodon* beds. Went out duck shooting after supper and shot a mallard and a muskrat.

**August 15** I got up very early before sunrise and walked to the lake and crept up to bag a shot at some wild geese. The lake was still and calm, occasionally a mallard would awake and give a warning quack. I fired at some ducks. Immediately there was an uproar. All its inhabitants awoke. The "yellow legs" screamed, ducks quacked, wild geese cahonked! and bayed like a pack of hounds. A couple of gulls come flying along to see what was the matter. Alarm was everywhere and the sacred stillness of the autumnal morning was broken. I came suddenly on a flock of geese and ducks. They rose with an uproar before I could get a shot. I killed a cinnamon teal.

August 16 Went with Reed and Ashley to *Sauranodon* No. 6 [not to be confused with quarry no. 6] and dug out sixteen feet of concretions containing a great number of consecutive vertebrae which averaged four inches in diameter also some ribs and five or six small rough bones apparently paddle bones. At the extreme end of our sixteen feet excavation, Reed found the humerus? of a paddle. We had to leave off after going twenty feet into the hill. We found belemnites and ammonites associated with the bones. We could not determine which end of the ten feet or so of bones is towards the tail or forwards the head. The paddle bones were found about six feet into the hill. There is little diminution of size in the vertebra from one end to the other. We packed home a heavy load of bones to the camp.[128]

August 17 Men shooting ducks.

August 18 Worked again at the Sauranodon Quarry No. 6 getting out the paddle bone and phalanges attached to the paddle. There were eight of them. A cross section of the concretion in which they were contained showed the presence of others also, the largest being nearest the humerus or limb bone of the paddle.

August 19 Worked at No. 12 at Robbers Roost. Got out a large flattened bone and some vertebrae. Made sketches. Men shot a wild goose.

August 20 & 21 At work at No. 12.

August 22 Worked at No. 2 south of station getting out a large vertebra thirteen to fifteen inches diameter and weighing one hundred pounds?[129]

August 23 The large bones in the *Laosaurus* horizon east of No. 3 are rotten. Some bones in the bed again below it between it and *Sauranodon*.

August 24 Lectured in the freight room on the form[ation] of the Earth etc.

August 25 & 26 No. 8 and No. 11. Packing up bones.[130]

A butcher bird (shrike) [loggerhead shrike] was feeding four young ones with great noise in an adjoining bush near the quarry. We saw the old bird pulling off the meat of something which proved to be a horned toad spitted lengthwise on a thorn.

Found signs of a hollow boned dinosaur in the upper portion of the bed above the Sauranodon horizon and intermediate between the *Sauranodon* and *Laosaurus* beds.

August 27 Worked at this last discovery [quarry no. 13] and got out the toe bone, claw and metatarsal of a carnivorous Dino; also some vertebrae six inches diameter.[131]

August 28–29–30 Continued my work alone.[132] The bones are in sandstone fifty feet above the *Sauranodon* beds.

August 31 Mrs. Chase had an oyster can full of young horned toads; ten of them pink in color and an inch in length.[133]

September 1 Prospected along my newly discovered ridge horizon. Found indications of bones at intervals for half a mile west of the station. The sandstone in places is beautifully ripplemarked; found a carnivorous tooth.

September 2 Went to vote at Medicine Bow.[134] The men rode on the cowcatcher. A half dozen houses, a store and saloon combined and a railway tank constituted the village of Medicine Bow.

September 3 Ran a trench along the new quarry.

A large squirrel hawk[135] alighted close to me without seeing me, I being partly hidden by an angle of the rock. He was so close that I could see his head vibrating to and fro as he watched eagerly some unseen prey on the opposite hill. Suddenly he launched forth across the ravine straight as an arrow and struck some small creature among the bushes. I could hear the crash of his wings as he struck the bush. Wonderful must be the sight of these birds for the, to me, invisible mouse was full fifty yards off and concealed by the bushes.

**September 4** Continued prospecting as far as No. 1. I found another horizon of bones in a sandstone thirty feet above the *Laosaurus* beds and at the base of the red and purple clays thus:

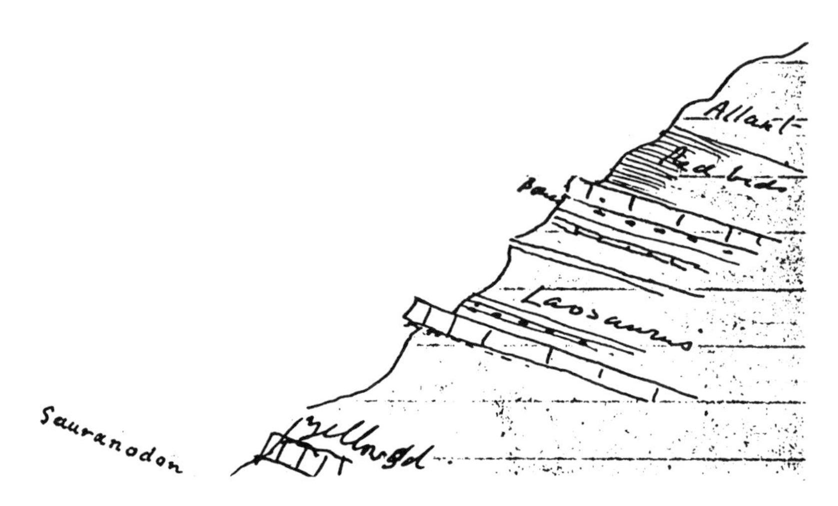


The shale that is quite thin above the *Laosaurus* beds becomes near No. 1, a thick massive sandstone conglomerative, below with impressions of herbage like reeds and fragments of bone. The sandstone is like that of *Titan[osaurus]*[136] at Morrison, the bones at the base are partly in sandstone, partly in shale.

**September** Spent most of the rest of the month at my 1½ quarry.[137] The weather has been beautiful and autumnal and the bushes soon began to turn a rich orange color.

My work has been solitary. My only company being a few large hawks. The life at the station has been enlivened by the arrival of the steel gang, fifty men in all who are daily seen dashing down the railroad track on their handcars.[138]

Reed and Cope's men have been preparing for the severities of the coming winter by making dug outs. These are little low cottages over a three feet deep excavation. The roof and sides are banked up with earth. The chinks and cracks are tinned over by flattened oyster cans. They are the warmed houses to withstand the fearful Wyoming blizzards. I shall stand in my tent as long as I can and then take refuge in the section house.

Towards the end of the month Marsh sent me to No. 12 to [near] Robbers Roost. A rather pretty spot but wild and lonely in the extreme situated amongst a series of ravines caused by a natural fold of the strata. Quarry 12 lies between two hog backs on a smooth hill of drab clay. We have opened a trench about fifty feet long, five deep and three wide. The bones found according to Marsh belong to a species *[ungulatus]* of *Stegosaurus* characterized by large plates of armor and long spines. The hind legs much longer than the fore legs. It is related to the *Stegosaurus armatus* we found at Morrison, Colorado.

October We often come upon broad flat plates of armor. . . . They are grooved and striated and show some internal hollow bone structures in little holes about the size of a lead pencil filled with calcspar [a naturally formed calcium carbonate]. Up to October 11th we got out about eleven bones consisting of vertebrae, ribs four feet long, toe bones, plates of armor, femur and humerus or part of the fore and hind legs.

In October about the 3rd, we heard of the fight with the Ute Indians on Milk River many miles west of us and of the death of Major Holdenburg [Thornburgh] and the Meeker Massacre.[139]

Troops kept frequently passing through in the trains and with them all the equipment of war. On the 3rd the whole northwestern horizon was enveloped in smoke leading us to suppose the Indians had fired the woods.[140] I had been obliged to work alone for some time at No. 12 but on the 6th engaged a boy named Jordain to help me. We got out only a few bones and portions of the feet.

October 14 On arriving at the Quarry, I found my broom carried fifty yards from the hole down a steep ravine by mountain rats as testified by their teeth marks. My butcher knife and ball of string were also missing. On climbing the ridge above my quarry I saw a buck and doe antelope in the ravine close below me. They bleated out their warning signal kerr-rr and trotted up the ravine.

October 15 Blew and snowed. Had to leave my tent and carry all my things into the section house which henceforth I made my headquarters with Mr. Ryan and wife and Bob.

October 16 Blowing a gale and snowing. No one able to go out.

October 17 Weather modulated. Quarry full of snow which during the winter gave us much trouble as it had to be thrown out before we could work at the bones.

Nathan C. Meeker, about 1878. Courtesy of the Colorado Historical Society.

## The Ute War

W. H. Reed's story concerning the North Park Indian scare during the summer of 1879 was but one in a number of such incidents. The 1870s had witnessed increasing pressure on the Ute Indians, who then occupied about twelve million acres of the western slope of the Rocky Mountains in western Colorado. Chief Ouray and other Ute leaders recognized the power of the whites who increasingly intruded on their traditional nomadic way of life, and they negotiated favorable treaties, after which their people began to farm. Other Utes resisted. They continued to hunt and, in addition, visited whites like Judge Costello for food.

In 1878 Nathan C. Meeker was appointed to the White River Indian Agency in northwestern Colorado. He was idealistic and strong-willed, and he had written a novel about the conversion of natives of a South Sea Island from barbarism to a utopian form of civilization. Meeker attempted to force the Ute band on the Milk River Reservation to adopt farming by reducing the pasture area for

Among the pests of our camp life whilst we were in the tent were hundreds of field mice with white bellies and soft tawny brown fur, black piercing beads of eyes and long tails.

As soon as the light was lit they made their appearance running,

Major Thomas Tipton Thornburgh. Courtesy of the Colorado Historical Society.

their horses and restricting the access to the food due them under treaty. When threats arose, Meeker requested support from the army.

Fearing the arrest of some of their leaders, on September 29, 1879, a band of Ute Indians ambushed Major Thomas Tipton Thornburgh and his troops on the Milk River in Colorado. Thornburgh, eleven other soldiers, and a teamster were killed. That same day another group attacked the Indian agency, killing Meeker and six others, burning the agency, and taking several women as prisoners. The three companies of cavalry were pinned down but fortified their position. Led by scout Joe Rankin, four of those besieged were able to escape, and riding 160 miles in twenty-seven and half hours, reached Rawlins on October 1. Three days later the remnants of Thornburgh's command were relieved by Company D of the Ninth Cavalry.

Chief Ouray arranged a truce before the arrival of hundreds of U.S. troops, who were set to suppress the Utes. The result of this outbreak was removal of the Utes from the entire twelve million acres of western Colorado to small reservations in Utah.

climbing and scrambling over everything. When the light was put out and I got into bed they literally swarmed and they could be heard gnawing at everything in the tent. As winter advanced their energies were directed to making nests preparatory for the winter. I had about a pound of cotton batting with me for packing. This they dragged out of my box and in the morning I found a cosy little nest made of it in the drawer where I kept my sketches. Sketches, papers, books, and envelopes being all enlisted in the structure whilst a copious supply of beans and rice had been hauled up bean by bean and grain by grain from my box of groceries for their winter commissariat. I soon rousted them out of this retreat but next day they had again got hold of the cotton and built a nest along side my bacon in the bacon box, an admirable plan for securing board and lodging for the winter.

I again took away the cotton and put it for safety under my pillow. This was worse than ever as they ran over my face and kept working at the cotton under the pillow and occasionally pulled at my hair perhaps in mistake.

Then they tore up my volume of Hayden's surveys and built another nest out of it and the cotton which I was forced to give up to them. They did an immense amount of damage. And in future I shall have everything shut up in tin boxes. One of their nightly amusements was to half fill my high boots and shoes with beans and rice.

One of the blessings of leaving my tent was future freedom from these pests.

Then on the night of the storm the siredon lizards appeared again, four of them marching solemnly into my tent. I put two of them in a vessel of water but they soon succumbed to the cold.

October 20 Weather fine resumed work getting out a carnivorous tooth one and half inches long and serrated and some other bones. Fired at two golden eagles on my way home. After supper the talk was about boxing and other feats of strength and about poaching in England.

October 21 Found another similar tooth.

October 22 [Found] a good toe bone core. I generally shot a cottontail or jack rabbit on my way to and fro. *Graphic* (English) arrived with my sketch in it.[141]

October 23 Shot a great butcher bird near the quarry and a snow white jack rabbit, the first I have seen this year in his winter dress.

October 24 After an unsuccessful day at [quarry no.] 12, I took my gun down to Rock Creek and fired at some small gulls and killed some suckers in the creek.

I found a newly made beaver dam and a beaver's hut. A beaver was sitting on the bank near it looking like a very large muskrat. I fired several times at the beavers but they ducked at the flash or else their thick hides warded off the shot. They seemed quite tame. It was about sunset and I stood on the top of the bank pouring down shot at them at twenty yards without any perceptible effect.

October 25 Spent the day sketching at No. 12. Then walked up the ravine and sketched the remarkable arch or fold in the strata. The center of this wide rather flat arch is much jointed. The stratum next above it on the south side pitches to the south but at a lower angle than the other end of the arch on the north. Up the ravine is a remarkable development of the massive cream colored sandstone above the arch showing excellent types of cross bedding. The upper part of this gently sloping sandstone is eroded into curious forms.

I lunched at the roost under a rock, a large crevice in which was entirely occupied by a mountain rats' nest. There was some fifteen feet of miscellaneous rubbish in the crevice consisting of sticks, leaves, bones of all kinds—amongst them the skull of a coyote, dung of coyote, features and parts of a hawk or eagle, spines of prickly pear, thorny branches of grease wood, [and] pieces of shale. The nest was big enough for a grizzly bear.

Their motive for this accumulation is doubtful, it may be for warmth or for protection.

Given a big hole cavern or crevice, they are bound to fill perhaps that there may be no room for any other creature to come in or to allow wind or snow from reaching them. This system of protection seemed to be implied by their carrying up bits of shale and prickly shrubs to defend the outside. The dung dropped by these animals accumulates in large masses in the crevice.

The Robbers Roost was a locality well chosen for the designs of the robbers. They and their horses were throughly screened from all view by the rocks and by a deep gash cut by water through the center of the ravine. Luxuriant grass grows along the bottom giving forage to their horses. In the center of the ravine grows a solitary large cottonwood which doubtless supplied them with fuel and shelter and from this retreat they could peep down the ravine and watch the advent of the train they intended to plunge into the ravine and then rob the coaches amid general disaster. After their hellish work they could easily have got away through the labyrinth of ravines which converge towards this spot.

The locality is a favorite resort of deer.

The geology at this point is very interesting.

The strata make a beautiful arch on the center of the fold. The arch is comparatively low and flattish. It is composed of massive gray soft sandstone much jointed and fractured by the tension of folding the Red Beds underneath the arch [which] appear nearly horizontal. The usual beds above it on the south are thicker than on the north side and the angle is less abrupt.

Whilst sketching I thought I heard voices and being rather nervous about Indians, I followed the ravine up into a dark corner out of which flew a long eared owl who doubtless had been muttering to himself. I followed up the ravine to the south and in a side ravine came suddenly on a couple of black tail does. They were as much startled as I was and stood for a moment starring at me within easy pistol shot. They bounded away with long graceful leaps.

October 27 I hired Bob Good to help me at the quarry [no. 12]. We got a few small bones and an armor plate. We went home together by the beaver dam. A beaver was swimming upstream and landed and began cropping leaves. I fired but he again escaped into the water.

Received a letter from Marsh telling me to go to No. 1A. Heard a coyote yap yap a hoo close to us.

October 28 Got out a fine plate two feet long by one [foot] wide. Brought in our tools from No. 12.

October 29 Went to work on Quarry No. 1A and got out a foot bone near end of tibia, probably of the fore foot; close to it a phalanx.[142]

October 30 We cut back the quarry for fifteen feet and found some vertebrae. Whilst digging we heard a shot below us and on looking down the hill saw a coyote running up the hill towards us. As he climbed the hill on the other side of the quarry I fired. In his course the coyote started a rabbit who fled with far greater terror from his deadly enemy than he would have from a man. The coyote had enough to do to take care of himself.

October 31 We carried our camp stove up to the quarry for warmth and for cooking our dinner. Found a large carnivorous tooth *[Allosaurus fragilis]*.

November 1 Occupied in sketching bones and packing and in making a bedstead of boards on the loft of the section house. Paid Ryan $10 for board for two weeks for self and Bob.

November 2 Found a carnivorous tooth.

November 3 Shot jack rabbits on way to Quarry 1A. They are now all snow white and in the evening jump out of the brush like ghosts or white cats.

November 4 Pitched my old tent in the quarry. Saw golden eagle near quarry. Shot another specimen of the great shrike. This is the last of the small birds remaining with us now.

November 5 Large biconcave caudal.

November 6 Blowing hard from the west and snowing.

November 7 Boxes and sketches of Robbers Roost strata [quarry no. 12].[143]

November 8 The month of November has been very variable in weather. Some days beautiful calm summer-like varied by days of snow and wind. We worked without much success at No. 1A finding the bones in a thin streak in shale below the sandstone. We found a small group of bones appearing to be the limb bones of some dinosaur no larger than a dog, the limb bones being about one foot long and some small toe bones from one to two inches long and five in number.[144] Got out from Skeleton 2[145] a large imperfect humerus fifty-six inches long. Also a tibia and fibula near humerus. For the fractured bones, we used plaster of paris. We found some hollow bones, toe bones two to three inches the inside filled with calcite.

Some of the hollow bones were flattened by pressure. The hollow bones, of which we found a good number, were very fragile and hard to get out perfect. I made a careful pen and ink sketch of the strata from Robbers Roost with No. 12 Quarry in foreground.[146] The arch is several hundred feet in span.

November 20 On the 20th we kept Thanksgiving Day by a holiday which I spent in the morning sketching Reed and Hubble's dugouts in the snow.[147]

In the afternoon Bob and I went out shooting to Robbers Roost on the hand car, B[ob] and I pumping with the men. We got out near the beaver dam and walked down the river over the ice. The top of the beaver's lodge just appeared above the ice. Here and there were clear smooth places in the ice made by the beavers as breathing holes.

The banks of the stream are cut into curious pillars. The clay nests of the cliff swallow adhered to the cliffs. Measured some strata at the Roost. Shot a sage rabbit. Reed shot a coyote.

*Winter Quarters at Como Bluff.* Reed is standing in front of the dugouts. Watercolor by Arthur Lakes from a sketch done Thanksgiving Day, November 20, 1879. Courtesy of the Peabody Museum of Natural History, Yale University.

December 8 Worked at No. 1A. Got out the rest of the large mass of cervical vertebrae but in a very rotten condition. Found a large carnivorous tooth three inches long. Near it we found more of the little hollow boned skeleton a claw core one inch long and a hollow toe bone close to it. Bob and I were caught in a violent snowstorm and reached the station blinded and half frozen.

Before the storm it was very mild and still but numbers of jack rabbits were running over the prairie looking for good shelter against the storm; they foresaw [what was] to be coming. A solitary gray crowned finch perched near the station with breast [of] magenta gray near the beak and crown of head black.

December 9 Abandoned Quarry 1A [Big Canyon Quarry] and went to 2B.[148]

January 1, 1880 All hands dined at the section house. After dinner we adjourned to shoot at a target with rifles. Reed made the best shot and I next. Weather summer-like.

January The weather in January was fine varied by cold snaps. Most of the time was spent at Quarry 12 *(Stegosaurus)* by Bob and I.

We discovered some enormous dermal spines which Bob called "devils tails." One was two feet long by six inches at the butt. We did not know what they were till we heard from Marsh.

We found a pair of large ones about the same size and two smaller ones. They were all near together amongst a string of caudal vertebrae. Also an ischium? [Yes] about thirty inches long another short spine eleven inches long and much flattened out towards the point. The latter was twenty feet from the other spines. We also found a long plate and some crocodilian scutes. Towards the end of the month we found another ischium five feet from the first and twenty-four inches long.

We ran a sort of short tunnel in toward the west end of the excavation.

Towards the end of the month the wall of the quarry began to crack and cave in.

We used to eat our dinner daily on a snow bank sometimes using the broad end of a geological pick as a spoon to eat our oysters with. Occasionally the weather was very cold from 20° to 30° below zero especially on the 30th with the pit half full of snow and fine snow blowing in upon us.

Ryan and Tom D. had a battle with a bull on a railway cutting. The bull refusing to let them pass. Tom felled it with a sledge hammer.

Muskrats came up from the marsh driven out by the ice and took refuge under the wood pile.

Gray crowned finches *(Leucostiste tephrocotis)* came in flocks about the station together with shore larks and snow buntings.

Bob left me on the 25th Jan. and I worked alone. Made sketch of *Sauranodon* quarry for Marsh. Game scarce.

February 2 The wall of quarry still holding but the crack widening.

February 3 Severe weather. Froze my ear. Train off track by cattle.

February 4 Very severe, blowing hard. Track men froze their ears.

February 5 Fine but very cold. A gale from NW. Packed by express a box containing supposed carnivorous skull and sacrum from Quarry 1A.[149]

Reed said that one day after heavy snow on passing near a snowdrift he found bushels of snow birds alive in it. They caught fifty of them. Snow buntings, *Junco hyemalis* [dark-eyed junco] and shore larks around station.

February 6 Blowing snowing.

February 7 Worked at [no.] 12.

February 8 Ryan saw three coyotes chasing a rabbit the latter ran so close to him in its terror he could have struck it with his hammer. The coyotes too came within stones throw of him finding themselves buffaloed of their prey, they sat up on their haunches and howled. The coyotes had separated and were trying to corral the rabbit.

February 9 Tapped a small spring at No. 12.

February 10 Hired a man called Richard Hallett with hair all down over his shoulders, a tall lean Yankee. We worked at [no.] 12 and got out two more spines. Reed shot an antelope. Expressed the spines to Marsh.[150]

February 11 Hallett and I fetched home the remains of the three antelope Reed had killed.

*The "Pleasures" of Science*. Self-portrait with Richard Hallett, digging out stegosaurus bones at quarry no. 12 during the Wyoming winter of 1880. Watercolor by Arthur Lakes. Courtesy of the Peabody Museum of Natural History, Yale University.

February 12 Very cold, snow blowing into quarry. R[eed] froze his foot, had to make a fire to thaw him.

February 13 Still cold, 15° below zero.

February 14 Too cold to work, blowing with blizzard. Reed told us about seeing a puma spring on a elk. The latter tried hard to throw him off. Reed waited till the puma had killed the elk and was sucking the blood from the throat. Then he whistled, the puma looked up Reed fired and killed him.

February 16 At [no.] 12. Got out a spine five feet below the ischium. It was right in the little pond formed by the spring. It was twenty-one inches long and four inches across butt end.

February 17 Found another spine four feet east of ischium lying upon a small flat armor plate as if crushed down upon it. The spine was very large being six inches at the butt, the top was missing as it abutted against another large and perfect spine behind it which was twenty-six inches long and six inches across the butt like a huge rhinoceros horn. The two together weighed four pounds. I expressed them to Marsh.[151]

February 18 Thermo 10° above freezing point.

February 19 Clearing out quarry.

February 20 Got out two large fine perfect caudals four inches diameter. We are troubled by water.

February 21 Digging holes to get rid of the water.

February 23 Beautiful and spring like. The first thaw we have had. Streams running everywhere.

The quarry's side fell in and left over a ton in the hole. We cleared it out and found portion of a small jaw perhaps crocodiles with teeth attached perhaps upper or lower jaw of a jaw already expressed to O.C.M.

February 24 Storming.

February 25 Troubled with water.

February 27 Very cold unable to go out.

Ryan saw numbers of jack rabbits running about stupefied by the cold. Reed says they make their forms on the eastside of a sage bush to avoid the prevalent northwest wind and when an east wind comes as on this occasion accompanied by snow, they are driven quite aback and fooled.

Bitterly cold at night, could not sleep from the cold although under a buffalo robe and two double blankets and quilt.

February 28 Fearfully cold. Thermo[meter] 30° below zero and blowing a blizzard and a perfect river of blinding snow. Clocks stopped by the cold. Water buckets frozen solid. Knife handles like icicle. No heat from the red hot stove. Track men dare not go out as it would have been fatal.

Went to Reed's dugout. Mrs. Chase had bought over her baby in the leather mail sack to Reed's dugout. Made diagram of Quarry 12 for Marsh who sent proof sheet of his article on *Stegosaurus*.[152]

March 1 A sparrowhawk flew over the quarry.

March 4 Very cold and stormy. Reed and I went out duck shooting. Ducks were plentiful. We shot a couple of pintail ducks. The pintail hibernates at big springs near the warm water.

We have worked all the week at the Roost and lunch there cooking our rabbits at the fire. The weather has been cold and storming. On the 12th it was 30° below zero. On Sunday the 13 two trains collided.[153] Though bars three inches thick were broken like reeds, yet the clock in the engine room was not stopped.

Fred B[154] said that there was a humming bird's nest in a tree in a logging camp and the lumbermen spared the tree till the young birds were fledged.

Engineer was killed on the train.

Reed told of a big pyramidal crystal of quartz eight feet long he found in a cave near North Park.

March 19 I left Como for Golden.

The strata along the U.P.R.R. [Union Pacific Railroad] for miles appears to be Cretaceous especially Cretaceous groups No. 2 and 3. A great deal of dark gray shale is exposed in the cuttings. The strata is uplifted into hogbacks dipping about 25° and towards the south.

At Miser is a deep railroad cut showing large cannon ball concretions in Cretaceous No. 3.

At Wyoming Station, the dip is toward the north.

As we approach Laramie we see Jelm Mountain to the west connected with the main range dipping north. There is probably a wide synclinal [a low troughlike area in bedrock where rocks incline together from opposite sides] between Jelm Mountain (Jurassic) and the Como hogbacks.

The sedimentary shale between the Laramie hills and the main range seems to have been wrinkled into several minor folds by lateral tangential forces exerted from the two ranges. The Cretaceous is thus brought to light at various places and at Wyoming probably the Laramie coal series. [See facsimile, p. 82].

The scenery along the track is the monotonous rolling prairie covered with snow with here and there a herd of antelope or some cattle. Every now and then we press through snow sheds. Miser, the short for misery, is a lonely miserable spot for a station. Another station is Lookout, Cooper Lake lies in a depression probably a fold.[155] Near Laramie City the strata begin to dish up toward the Laramie hills.

The city is the usual prairie town, its strength being the machine and rolling mills. I stopped at the works. Old rails are cut into blocks by a huge shears then welded into flat pieces about five feet long, heated red hot and passed and repassed through a series of grooved rollers till they come out long snakelike rails glowing with red heat.

Then they are passed on rollers to a saw which cuts them the right length. From the rolling mill[156] Mr. W[anless] and I went [to] the machine shop where the wreck of an engine was being repaired. Thence to the roundhouse were the great ironhorses are stabled and groomed and got ready to start.

Evening, visited the Wanless's.

I found clean sheets and a civilized bed a great treat after my camp life.

On leaving Laramie we pass in a cutting through red Triassic rock dipping gently West. Soon we come to course red granite at Sherman.[157] This is the axis of the main range.

Beyond this in [a] granite canyon the Trias is seen reclining at a steep angle directly upon the granite being the other side of the Laramie upheaval. Further on is a greenish schist. Once there, the prairie strata is horizontal.

## Some Geological Notes on the Como Region

Following the railway track to the west of Como station we pass through a railway cutting where the Fort Benton, the Dakotah Cretaceous, and the Atlantosaurus Jurassic beds are exposed down to Triassic. This is on the North shore of the lake. The strata in the cutting dip about 70° unto the South, further on along the railroad track we come to the dark shales of the Fort Benton group dipping at various angles 70°, 45°, 60°, and 80°; the average is between 60° and 75° and to the South.

Towards the bridge at Robbers Roost the strata begins to curve round towards the west end of the fold. The angle of inclination of the Dakotah rocks is nearly vertical with a dip of about 10° to the South. Inside of the Dakotah we have a fine exposure of the *Atlantosaurus* and *Sauranodon* beds and the Trias which now begin to dip to the North at an angle of about 20°. Following the curve of the strata to its limit at the west we find the strata rapidly curving around till it becomes nearly horizontal or not more that 10 to 15 degrees.

Going West we pass over the several semicircular outcrops like the tiers of an amphitheatre from the Dakotah to the Fort Benton lying back of it and behind that the wave-like crescent of the Fort Pierre group apparently completing the outer margin of the anticlinal curve or fold.

The outer rings of the anticlinal are of course the widest apart and narrow as we approach the inner crescent the fold flattens down toward the center to a dip of not more than 10° and dips gradually North and South. The heavy massive cream colored sandstone of the upper member of the Trias or basal member of the Jurassic forms a prominent outline. The inner portion of the fold is composed of the Red Trias sandstone.

The ridges to the North have the steepest dip. The Red Trias beds appear at the outer edge of the anticlinal but are grassed over by the Como basin on the North side. In the center of the basin or rather to the South side lies Lake Como.

The ridges pass up toward the East where the red beds are again exposed, flanked on the northeast by the Dakotah group and Jurassic beds. Thus at Como we have an oval basin excavated out of an anticlinal fold about two miles wide by about six miles long.

The strata at the extreme west end of the fold all curve in like the walls of an amphitheatre only inclining a good deal inwards towards the basin. The side walls being at the north side only a little below the vertical and at the south side inclining to the south at an angle of 25°.

No. 2 Quarry is 100 feet from the top of the bluff or from the base of the Dakotah sandstone. So also is the spot where crocodiles teeth were found. This is the point on the bluffs due south of the railroad tank.

In a prospect hole fifty paces west of No. 2 are some large limb bone fragments.

Color of the bed of variegated shales:

base—salmon red, purple bands, and light gray bands

top—brown rough iron concretions often falling on the floor of the trough.

Above this is a rounded portion of the bluff of light-gray shale traversed by rows of concretions. Under the top most of these rows of concretions is a whitest streak in which crocodile remains are common.

There is a good section of the bluff about one mile East (W.?) of

## Rough Section Due South of the Station

| | Thickness in feet |
|---|---|
| Beginning at the base and ascending upwards | |
| Alluvium and clay | |
| A gulch revials (?) | |
| *Triassic Red Beds* | |
| First soft yellow massive sandstone and shale nearly horizontal | 10 |
| Dark red shaley sandstone dipping slightly south mottled with light gray | 10 |
| This makes bench or ridge No. 1 | |
| Brick red shale capped with light gray slabs dipping 20° south 20 forming bench No. 2 | |
| Red shale | 15 |
| Massive red sandstone | 10 |
| Brownish red sandstone | 10 |
| Bench No. 3 dip 25° South | |
| Massive cream colored sandstone | 25 |
| Reddish sandstone | 10 |
| Reddish shale | 10 |
| Gray shale | 15 |
| Coarse gray shaley sandstone | 10 |
| *Jurassic* | |
| Very massive cream colored sandstone forming the base of the Jurassic | 80 to 100 |
| Bluff of 100 feet covered apparently consisting of | 100 |
| Greenish gray shales | |
| Red band near top | |
| Gray shales | |
| Shaley coarse whitish gray sandstone | |
| *Sauranodon beds* | |
| Greenish-gray shales and their shaley limestone containing cubes of iron pyrites and belemnites | 60 to 80 |
| *Atlantosaurus beds* | |
| Yellow sandstone | 45 |
| Light gray shale capped with five feet of brown flagstones | |
| Light gray salmon red and purple shales capped with ten feet of brown shaley sandstone | 80 |
| Purple clays and shales and reddish | 100 |
| This forms the top bench | |
| The bench over this bench consists of purple and gray shales with crocodiles' teeth | 60 |
| Gray clays and shales | 150 |
| Black sandy shale | |
| *Cretaceous* | |
| Dakotah sandstone forming the top of the bluff | 50 to 100 |

Como station about five hundred yards west of this is Sauranodon No. 2 Quarry about four feet below the sulphur yellow sandstone and near to the top of the middle bench in the bluff.

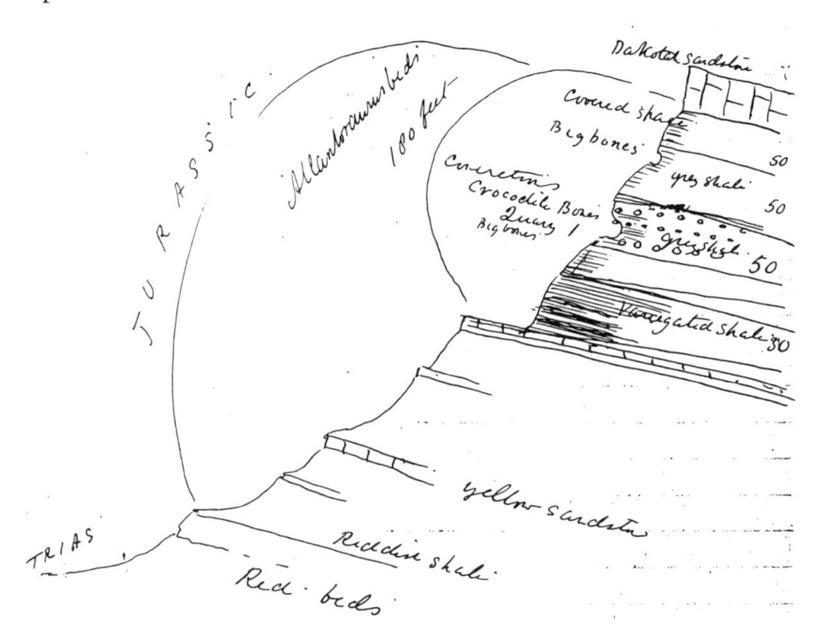


North of Como Station. The north shore of Lake Como is bounded by a cliff about twenty feet high of yellow cream colored sandstone underlaid by patches of red course sandstone and then by more cream colored massive sandstone soft and twenty feet thick. This is overlaid by red shales.

This represents the bench below the *Sauranodon* beds well displayed on the opposite bluff south of the station and just above the massive cream colored sandstone of the Trias or lowest Jurassic.

The west side of the lake is of a bank of lacustrine clays twelve feet thick on high above the water.

In the southwest end of the lake the red beds appear again.

Above the north bank of [the] lake is a jumble slope occupied by *Sauranodon* beds with belemnites. Then two low ridges of nearly vertical limestone dipping slightly north representing the bed above the yellow sulphur colored sandstone with its ripple marks. These outcrops form a low wall, below them is a trough occupied by purple and variegated shales, below the *Atlantosaurus* beds; and above these the bluff contains the *Atlantosaurus* beds and capped by the Dakotah group. The *Atlanto* beds are

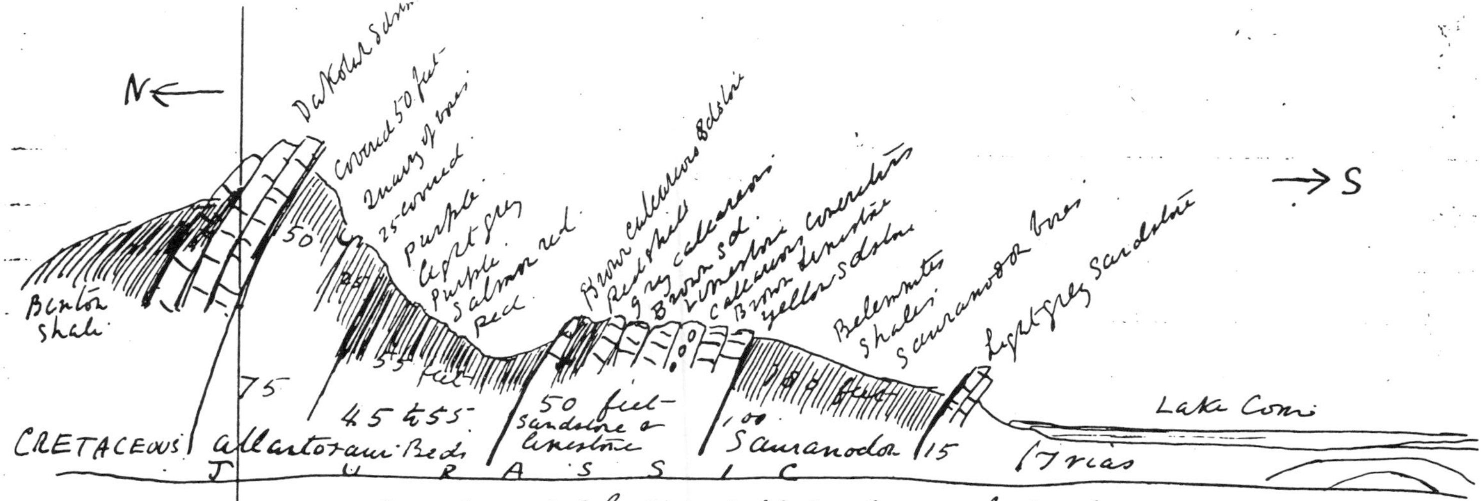

N ←

Dakotah Sdstn

Covered 50 feet

Quarry of bones

25 covered

purple

Light grey

Purple

Salmon red

Red

Brown calcareous Sdstone

Red shell

grey calcareous

Brown sdst

Limestone concretions

Calcareous

Brown limestone

yellow sdstone

Belemnites

Shales

Sauranodon bones

Light grey Sandstone

→ S

Benton Shale

50

25

75

55 feet

45 to 55

50 feet sandstone & limestone

100 feet

100

15

Lake Como

CRETACEOUS | Atlantosaur Beds | Sauranodon | Trias

J U R A S S I C

Section of bluffs on North Shore of Lake Como

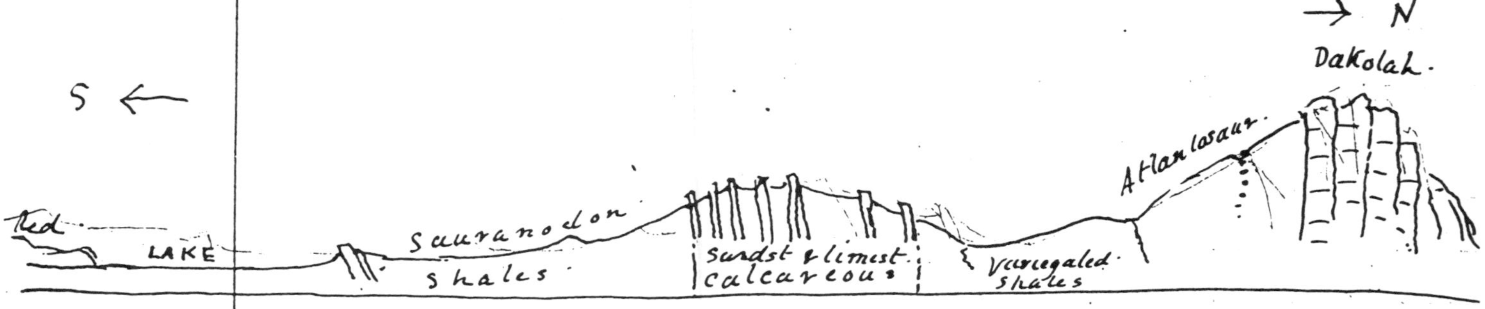

S ←

→ N

Dakotah

Atlantosaur

Red

LAKE

Sauranodon

Shales

Sandst & limest calcareous

Variegated Shales

Profile on N. Shore of Lake

Section of Dakotah and *Atlantosaurus* Beds on Railroad Track One Mile West of Como, West End of Lake

| | Feet Thick |
|---|---|
| Cretaceous | |
| Dakotah shales and sandstones | 150 |
| Black shale with silicified wood | 5 |
| *Atlantosaurus* beds | |
| Ash colored and light greenish shale | 75 |
| Purple and violet shales shading down into dark red with greenish purple bands at intervals | 85 |
| *Sauranodon* beds | |
| Also some layers of shale calcareous sandstones and soft yellow sandstone and gray shale | 100 |
| Soft yellow sandstone | 6 |
| Sandy shale | 30 |
| Trias | |
| Massive cream colored Jura Trias sandstone | 50 |
| Soft light gray sandstone | 20 |
| Red Triassic sandstone | |

about two hundred feet thick or one hundred feet [of] a vertical bed of coarse blackish sandstone separates the Dakotah from the Fort Benton clays and shales.

## View from Top of Dakotah Ridge above *Atlantosaurus* Beds

To the north east in the distance are a series of uplifted hogbacks or benches forming Freezeout Mountain. They appear to be Jurassic hogbacks.

East of this, the same beds dipping to the west are uptilled by the granite uplift of the Laramie range of black hills. Thus a trough or wide synclinal must be formed between the region of Freezeout and the Laramie hills. The center of the trough running north & south.

Nearer us are bluffs of the Jurassic and red beds dipping to the south at an angle of 25° and the interval of about four miles width is excavated out forming the basin fault of which at the lower western end is occupied by Lake Como.

At the base of the Dakotah cliff and *Atlantosaurus* beds, the lower portion of the *Atlantosaurus* beds are eroded back and stand out as a series of pink white and gray rolling hillocks cut by numerous little ravines acting as low foothills so to speak to the main *Atlantosaurus* and Dakotah cliff.

These hillocks or low mounds are bounded on the north by a low ridge or bench of the Red Beds forming the edge of a trough at an angle of 25°.

## Valley of Rock Creek

The Dakotah and Red Beds about three miles east of Lake Como begin to curl around to form the end of the ellipse at a point where Rock Creek passes through the strata called Rock Creek Gap. There is a remarkable display of strata tilted at a steep angle of 45° or more. The whole series of Dakotah rocks and red beds are shown bending inwards and dipping to the north. The Dakotah is tilted at the steepest angle; the inner strata are less vertical and from 45° gradually lessen down in steepness till at the top of the mesa about 250 feet high above the stream and one quarter mile from the Dakotah, the stratum becomes nearly horizontal forming the top of the fold or arch. It then sweeps rapidly around the basin occupied by Rock Creek.

Passing from this Rock Creek Gap around to Quarry No. 3, we soon reach nearly horizontal strata while the general dip is to the SE showing that we are on the other slope of the arch.

Crossing a flat park where there are powerful bubbling springs, we come to the first stratum forming the southern slope of the arch. It is a coarse heavy bedded white sandstone dipping 15° to the south and next above it is a red sandstone.

The bottom of the valley is occupied by the red beds. The slope to the north of the red beds is much steeper than the corresponding slope to the south. The river divides the two series, part of the floor of the valley being occupied by the lower *Sauranodon* and belemnite beds of the Jurassic. The "snail" shells found by Marsh are in argillocalcareous [composed of clay and calcium] gray concretions about 20 feet above the *Atlantosaurus* quarry.

Following the stream toward the SE through a sloping valley we find the east side of the valley to be formed of the top stratum of the *Sauranodon* beds. Whilst as we go on up the valley the lower portion of the *Atlantosaurus* beds from which the Dakotah cap has been removed by erosion lies a wide pink and greenish flat.

Rock Creek takes a sudden bend and cuts through the Dakotah group, Dakotah rocks forming the bank on either side. To the south No. 2 and 3 Cretaceous with their dark shales form the rounded banks of the stream.

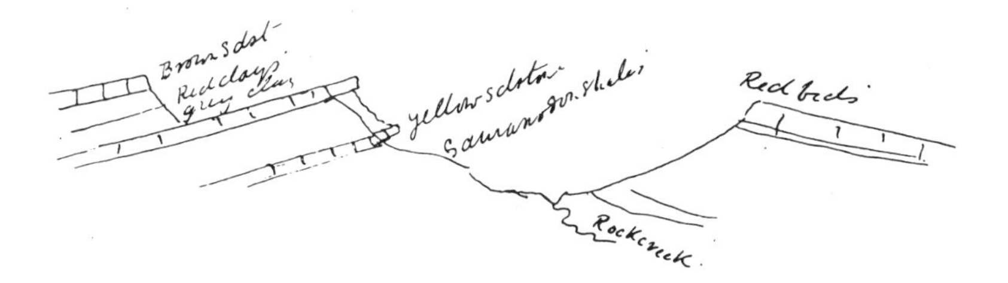


Quarry No.4 lies on the lower bone bed only ten feet above the variegated red and purple bed and 150 feet below the Dakotah ridge. About fifty yards north of it is a small quarry at an horizon 25 feet higher than No. 4 showing some large bones. No. 4 appears to correspond with what we have called *Morosaurus* [camarasaurus] horizon or No. 2. About forty feet below the Dakotah ridge and five hundred yards south of No. 4 is another prospect hole in a much higher horizon showing also a few large bones.[158]

Two hundred yards from this last or 800 yards from No. 4 close to the ravine where Marsh startled the elk at which Reed fired, is another quarry at a little higher horizon than No. 4 called Snail Shell Quarry from the shells Marsh found there.[159]

In this quarry is the rather small sacrum of a carnivorous saurian probably an "*Allosaur*"? This quarry is eighty feet below the Dakotah.

These quarries were excavated by Reed and Williston Jr. in the banks of Rock Creek. The valley of the creek is about two hundred feet deep and one half mile wide. It is bounded on the south by the *Atlantosaurus* and *Sauranodon* beds, on the north by a coarse, durable thickness and display of nearly horizontal red Triassic beds. The angle of the dip of these beds is from 15 to 20° dipping to the South.

At the east end of the Como basin, the strata curl around forming the end of the elliptical fold. The Dakotah forms a prominent ridge rudely fortified by Indians and commanding a view of the entire basin. The strata are here at a very small angle and dips due east. "Two-Bow Bone" quarry is thirty-four feet above the purple beds along the bluff and about fifteen feet below the crocodile horizon.[160] *Moros[aurus]* or Quarry 3 is twenty feet above the red and purple beds.

From Castle Rock going west a broad trench lies between the Dakotah and *Atlantosaurus* beds filled by the eroded strata of the purple and red beds. The north rim of the trench is made by a prominent bench of the *Sauranodon* sandstone (yellow schist), thus

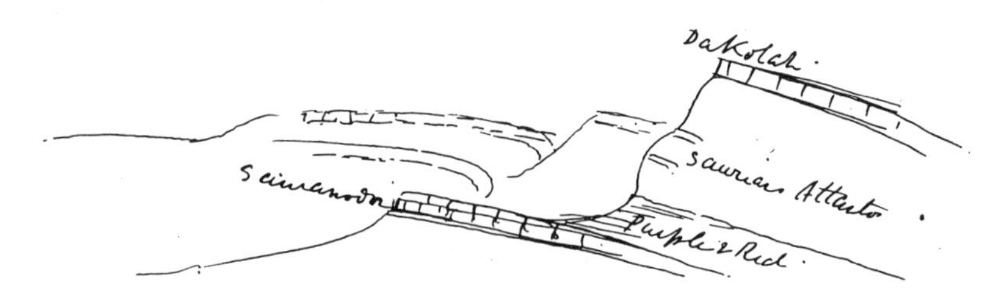


The interval is filled with rolling mounds of the eroded purple red and greenish shales at the base of the *Atlantosaurus* beds amongst them a few outcrops of harder concretionary material, fragments of bones washed from the beds above, masses or blocks of brown agate or chalcedony, selenite flakes, fragments of a very hard flinty dark olive green concretion.

From the Quarry No. 3 across the basin to Rock Creek is about three miles. The Union Pacific Railroad track passes east and west through the center of the basin. The clay and shaley slopes of the bluffs are more or less destitute of vegetation.[161]

# Appendix: History and Synonymy of Some Jurassic Dinosaur Species

In the spring of 1877, Arthur Lakes sent O. C. Marsh of Yale two and half sacral vertebrae of a large dinosaur (acc. no. 968 at Yale's Peabody Museum) found in what was later designated quarry no. 1 near Morrison, Colorado. In July, Marsh described and named this specimen *Titanosaurus montanus,* which was catalogued YPM 1835. Lakes then wrote to Marsh, asking him if he was interested in a second specimen, consisting of several vertebrae that he had collected at quarry no. 10 near Morrison. Marsh was slow to respond, so Lakes sent the vertebrae for identification to Edward Drinker Cope in Philadelphia. Marsh learned of this development and quickly wrote to Lakes offering to buy this specimen. Lakes then wrote to Cope, asking him to send the vertebrae to Marsh in New Haven (acc. no. 993). To assist Lakes in his collecting at Morrison, Marsh dispatched Benjamin Franklin Mudge, who soon brought in his student Samuel Wendell Williston. Later in 1877 Marsh described specimens sent from Morrison quarry no. 5 as *Stegosaurus armatus* (YPM 1850). Included was a set of sauropod teeth that was subsequently assigned (in 1884) to diplodocus as *Diplodocus lacustris* (YPM 1922).

In August 1877, Mudge went to Cañon City, Colorado, in order to purchase bones from Oramel A. Lucas but they had already been purchased by Cope, who then employed Lucas. Cope described them as *Camarasaurus supremus* (later catalogued AMNH 5760). Mudge then opened another quarry southeast of where Lucas was digging, and he brought Williston for assistance (Garden Park quarry no. 1). They collected some fragmentary bones of a theropod later described as *Allosaurus fragilis* (YPM 1930) and a tail and hind limb of a sauropod described in 1878 as *Diplodocus longus* (YPM 1920).

William Harlow Reed and William Edward Carlin had opened quarry no. 1 at Como Bluff, Wyoming, at about the same time as had Lakes in the spring of 1877. They offered specimens to Marsh, who accepted the offer and sent Williston from Cañon City to Como Bluff to help them. All of this activity culminated in December with Marsh's description of the material at Morrison quarry no. 10 as *Apatosaurus ajax* (YPM 1860) and the material from Como Bluff quarry no. 1 as *Apatosaurus grandis* (YPM 1901).

Early in 1878 Marsh published an article describing *Apatosaurus ajax*.

Summary Table of the Synonymies

| Name | Status |
|---|---|
| Genus | |
| *Apatosaurus* Marsh | valid |
| *Atlantosaurus* Marsh | indeterminate |
| *Brontosaurus* Marsh | = *Apatosaurus* |
| *Camarasaurus* Cope | valid |
| *Morosaurus* Marsh | = *Camarasaurus* |
| *Stegosaurus* Marsh | valid |
| *Titanosaurus* Marsh (non Lydekker) | indeterminate |
| Species | |
| *Allosaurus fragilis* | valid |
| *Allosaurus lucaris* | = *Allosaurus fragilis* |
| *Apatosaurus ajax* | valid |
| *Apatosaurus laticollis* | = *Apatosaurus ajax* |
| *Atlantosaurus immanis* | = *Apatosaurus ajax* |
| *Atlantosaurus* (originally *Titanosaurus montanus*) | indeterminate |
| *Brontosaurus amplus* | *Apatosaurus excelsus* |
| *Brontosaurus excelsus* | *Apatosaurus excelsus* |
| *Camarasaurus supremus* | valid |
| *Camptosaurus (Camptonotus) amplus* | valid |
| *Diplodocus lacustris* | valid(?) |
| *Diplodocus longus* | valid |
| *Labrosaurus ferox* | *Allosaurus fragilis* |
| *Labrosaurus lucaris* | *Allosaurus fragilis* |
| *Laosaurus consors* | *Dryosaurus altus* |
| *Morosaurus grandis* | *Camarasaurus grandis* |
| *Morosaurus impar* | *Camarasaurus grandis* |
| *Stegosaurus armatus* | valid |
| *Stegosaurus ungulatus* | valid |

In the same paper, realizing that this genus *Titanosaurus* had been previously used a few months earlier by Lydekker for a related Indian sauropod dinosaur, Marsh changed its name to *Atlantosaurus montanus*. He also described a second skeleton from Morrison quarry no. 10 as a second species of atlantosaurus, named *Atlantosaurus immanis* (YPM 1840). The specimen called *Atlantosaurus montanus* is now considered indeterminate because of its poor condition, and *Atlantosaurus immanis* has become a junior synonym of *Apatosaurus ajax*. A few months later Marsh described a second specimen (a sacrum) from Garden Park quarry no. 1 as *Morosaurus impar*. He transferred his *Apatosaurus grandis* to this genus as *Morosaurus grandis*. It is now realized that the genus *Morosaurus* Marsh is a junior synonym of *Camarasaurus* Cope and that the two species, *Morosaurus impar* and *Morosaurus grandis,* are also synonymous. Since the latter has priority, it is now known as *Camarasaurus grandis* (Marsh). Late in 1878 Marsh described and named his *Diplodocus longus* (YPM 1920) from Garden Park quarry no. 1. Marsh described a second species of allosaurus, *Allosaurus lucaris,* from Como Bluff quarry no. 3.

In 1879 Marsh added a second species of stegosaurus, *Stegosaurus ungulatus,* from Como quarry no. 12 (YPM 1853). He described a big cervical vertebra collected in 1877 by Lakes and Mudge as *Apatosaurus laticollis* (YPM 1861). This was later found to be part of the skeleton of "*Atlantosaurus immanis*" (YPM 1840) and is thus referable to *Apatosaurus ajax*. Later in the year, Marsh described a sacrum and some vertebrae collected by Reed, Lakes, and Ashley at Como quarry no. 10 as *Brontosaurus excelsus* (YPM 1980). During the ensuing year much more of this skeleton was collected and its mounted skeleton can now be seen at the Yale Peabody Museum. *Allosaurus lucaris* was assigned to a new genus as *Labrosaurus lucaris*.

A specimen collected by Reed, Ashley, and Lakes at Como quarry no. 11 was named by Marsh in 1881 *Brontosaurus amplus* (YPM 1981). Today it is considered to be a junior synonym of *Brontosaurus excelsus*. In 1884 Marsh described *Diplodocus longus* more fully, referred the teeth of *Stegosaurus armatus* to *Diplodocus lacustris,* and he referred a second species to *Labrosaurus,* as *Labrosaurus ferox*.

In 1903 E. S. Riggs wrote the definitive paper about apatosaurus and showed that *Brontosaurus* is a junior synonym of *Apatosaurus*. In about 1914, Henry Fairfield Osburn and C. C. Mook showed that *Morosaurus* Marsh was a junior synonym of *Camarasaurus* Cope.

# Notes

## Preface

1. Smithsonian Institution Library Accession Book: February 24, 1898—February 6, 1903, nos. 160221–172766, uncatalogued volume, Special Collections Department, Smithsonian Institution Library.

## Introduction

1. J. McKeen Cattell, "Professor Arthur Lakes," *American Men of Science,* 2d ed. (New York: Science Press, 1910), p. 268; "Arthur Lakes, Sr.," *Mining Science* 68 (August 1913): 73–74; *Battell's Book,* 1862–63, and *Cautions Book,* Queens College Archives, Oxford University.
2. Allen DuPont Breck, *The Episcopal Church in Colorado, 1860–1963* (Denver: Big Mountain Press, 1963), pp. 60, 85, 310, 367, 371, and 377.
3. Arthur Lakes, "The Dinosaurs of the Rocky Mountains," *Kansas City Review,* 2 1879, pp. 731–35.

## Journal of Travel and Exploration: Discovery of Dinosaurs at Morrison, 1878

1. Lakes had conducted services at Idaho Springs, the seat for Clear Creek County, which is located in the small valley known as Bergen Park. The route from Golden to Morrison follows Bear Creek.
2. Henry C. Beckwith (1839–85) was a retired naval officer who obtained the rank of first assistant engineer having served aboard a number of vessels, including the *New Ironsides* from 1862 to 1876. Lakes was searching for fossil plants for Leo Lesquereux (1806–89), whose research was published as part of the Hayden Survey's reports.

3. These three periods constitute the Mesozoic Era, when dinosaurs existed: the Triassic Period spanned 224–208 million years, the Jurassic Period 208–145 million years, and the Cretaceous Period 145–65 million years. These dates are approximate and are from W. Brian Harland et al., *A Geologic Time Scale 1989* (Cambridge: Cambridge University Press, 1990).
4. Alexander S. Young was the agent (superintendent) for the Morrison Stone and Lime Company. 1877 *Colorado Directory*.
5. Dr. J. C. Donham is identified as the physician for Morrison in the 1875 *Colorado Directory*.
6. This bone was described by O. C. Marsh in "Notice of a New Gigantic Dinosaur," *American Journal of Science*, 3d ser., 14 (1879): 87–88. He first named the creature *Titanosaursus montanus* and later *Atlantosaurus montanus*. YPM 1835, sacral vertebrae. This specimen is indeterminate.
7. Lakes's first letter to Marsh, April 2, 1877, described a "Herculean war club" that has never been identified among the specimens at the Yale Peabody Museum. *Othniel Charles Marsh Papers*, reel 10, frame 355, Sterling Library Special Collections, Yale University.
8. The Dakotah Group, also known as Cretaceous no. 1, was one of the strata levels identified in the Hayden surveys analogous to the Lower Cretaceous period. Lakes makes mention of the 1874 Hayden Survey.
9. During the Cretaceous there was an inland sea covering much of the Rocky Mountain and Plains area from the Arctic Ocean to the Gulf of Mexico.
10. G. Kruse and G. W. Charles were listed as blacksmiths in Morrison. 1877 *Colorado Directory*.
11. J. G. Pease was proprietor of the Morrison House hotel in the 1875 *Colorado Directory*. He is listed as a miner in the 1880 census but had numerous boarders.
12. March 28, 1877. This phrase was crossed out.
13. Lakes corresponded with and sent specimens to Othniel Charles Marsh and Edward Drinker Cope, who identified them and became interested in this site. See Lakes correspondence with Marsh during the spring and summer of 1877, *Marsh Papers*, reel 10, frames 349–473. There is no record of the turtle bones at Yale. This reference indicates that these incidents were described sometime after they occurred.
14. The words "we discovered" are crossed through here.
15. It was from Yale site Saurian quarry no. 2. Unidentified segments of lower end of sauropod femurs. They arrived as part of YPM acc. no. 968 and, as of 1996, these bones have not been catalogued.
16. Lakes to Marsh, April 2, 1877, *Marsh Papers*, reel 10, frames 355–56. YPM acc. no. 968.
17. The following sentence is crossed out at this point: "Having engaged the help of two of my former pupils I bought a tent."

18. This is a reference to the discovery of Yale Saurian quarry no. 10, now known as Morrison quarry no. 10, from which *Apatosaurus ajax* was excavated.
19. The word "(Fig.)" appears here in the journal, but there is no illustration.
20. They were students at Jarvis Hall and the School of Mines. George Lyman Cannon Jr. (b. 1860), went on to have a distinguished career as a scientist and teacher at the Denver High School. He wrote a number of articles and books on science. See *Who's Who in America: A Biographical Dictionary of Notable Living Men and Women of the United States, 1901–1902,* ed. John W. Leonard (Chicago: Marquis, 1901), p. 177. Cannon briefly described his experiences working on this dig in "The Geology of Denver and Vicinity: Address of the Retiring President," *Proceedings of the Colorado Scientific Society* 4 (1891, 1892, 1893), pp. 244–46. A Thomas Elliott, age 24, is listed in the 1880 Colorado census as living in Manitou Springs and working as a liveryman and guide in the mountains.
21. The Connecticut River Valley has numerous footprints that Rev. Edward Hitchcock described as birdlike creatures. See, e.g., "An attempt to discriminate and describe the animals that made the fossil footmarks of the United States, and especially of New England," *Transactions of American Academy of Arts and Science,* n.s., 3 (1848): 129–259.
22. Several native American cultures, including the Plains, Woodland, and Ute, used the hogbacks for a variety of purposes over a number of centuries. See Kevin D. Black, *Archaeology of the Dinosaur Ridge Area* (Denver: Friends of Dinosaur Ridge and the Colorado Historical Society, with the Colorado Archaeological Society and the Morrison Natural History Museum, 1994).
23. The Upper Jurassic was approximately 157–145 million years ago and the Lower Cretaceous approximately 145–97 million years ago; see *A Geologic Time Scale 1989*. For a modern analysis of the geology of the Morrison area, see L. W. LeRoy and R. J. Weimer, *Geology of the Interstate 70 Road Cut Jefferson County, Colorado* (Golden: Department of Geology, Colorado School of Mines, 1971).
24. Morrison was on the route of the Denver, South Park, and Pacific Railroad that would not ascend through Kenosha Pass into South Park until 1879. See 1877 *Colorado Directory*; Virginia McConnell Simmons, *Bayou Salado: The Story of South Park,* rev. ed. (Boulder, Colo.: Fred Pruett Books, 1992), 169–71.
25. The teeth are those of the carnosaur *Allosaurus fragilis;* the larger creature is the sauropod, *Apatosaurus ajax*.
26. Lakes and his crew moved after this accident from Saurian quarry no. 10 to Saurian quarries nos. 1 and 4. See Lakes's letter to Marsh, June 15, 1877, *Marsh Papers,* reel 10, frames 400–402.
27. W. S. Smith ran a general merchandise store in Morrison in 1877. He had been president of the school board in 1875. See 1875 and 1878 *Colorado Directory*.

28. YPM acc. nos. 1046 and 1047.
29. Again, this reference indicates that Lakes wrote this journal sometime after the events described because the identification of the fossils only occurred after they had been shipped back east by Marsh and his staff. They are not identified among the Yale collections.
30. Yale quarry no. 1, which was prospected during the spring of 1877.
31. George Cannon worked at what became known as quarry no. 8, while Lakes and Thomas Elliot continued work at quarry no. 5.
32. This find, part of YPM 1850, included the teeth of *Diplodocus lacustris,* later given YPM 1922, as well as the type specimen for *Stegosaurus armatus* as well as unidentified sauropod limb bones. See "New Order of Extinct Reptilia *(Stegosauria)* from the Jurassic of the Rocky Mountains," *American Journal of Science,* 3d ser., 14 (1877): 513–14.
33. In Numbers 13:1–27 and Deuteronomy 1:19–25, the twelve scouts whom Moses was directed to send into the land of Canaan spied upon the lands and cut a cluster of grapes and other fruit in a gorge called Eschol near Hebrum and returned triumphantly to the Israelites.
34. Quarry no. 8 produced specimens for *Apatosaurus ajax* (YPM 4676).
35. Lakes conducted services at a number of frontier communities and mining camps intermittently during the 1870s. See Allen DuPont Breck, *The Episcopal Church in Colorado, 1860–1963* (Denver: Big Mountain Press, 1963), pp. 60, 85, 310, 367, 371, and 377.
36. Calvan Camp of the Chicago Creek district of Clear Creek County is listed as a miner in the 1880 Colorado census.
37. Bishop John Franklin Spalding (1828–1902) served as missionary bishop of the Episcopal Church in Colorado and Wyoming from 1874 to 1902. His wife was Lavinia Deborah Spencer Spalding and among his five children, his eldest son, Franklin Spencer Spalding (1865–1914), also became an Episcopal bishop in the western United States.
38. F. E. Everett was the proprietor of the Everett Bank in Golden.
39. Lakes wrote that June 19, 1877, was a Monday, but it was a Tuesday. Lakes's prior entry for June 18 likely occurred on Sunday, June 17, 1877.
40. L. M. Shields was listed as working as an engineer in Golden in the 1880 Colorado census.
41. These bones are not identified among the collections at the Peabody Museum although they are catalogued as part of YPM 1850, and some of them do not belong to *Stegosaurus armatus*.
42. Jarvis Hall was the Episcopal prep school in Golden at which Lakes taught a number of subjects, including composition and later geology, from its beginning in 1869. The spring semester continued until close to the Fourth of July during the 1870s.

43. *Atlantosaurus montanus* (YPM 1835). It is indeterminate and still bears that designation since further identification is impossible because of its condition.
44. *Goniopholis felise* has priority as the name for this crocodilian species.
45. Benjamin Franklin Mudge (1817–79) had been a professor at the Kansas State Agricultural College in Manhattan and worked as a collector for Marsh during the 1870s. He was responsible for the discovery of numerous important marine, avian, and terrestrial fossils from the Mesozoic in Kansas. See p. 26.
46. The word "men" is crossed out.
47. This incident apparently occurred during the summer of 1875; see *The Life of Benjamin Franklin Mudge* in Kansas 1861–1879, comp. Melville and Dorothy Mudge (Lakewood, Colo.: privately published, 1990), pp. 230–31. The Sioux were reported to have raided a group of cowboys and their ponies the next day.
48. These included Harry A. Brous, E. W. Field, and Samuel W. Williston.
49. They include *Hesperonis regalis, Ichthyornis dispar,* and several pterodactyls.
50. This incident occurred in May 1876 along Butte Creek in Kansas. See Life of Mudge, pp. 159–60.
51. The tune is probably "Tis Dawn the Lark Is Singing," lyrics anonymous, music by George James Webb, 1837. This tune was later used for the hymn "Stand Up, Stand Up for Jesus." *The Great Song Thesaurus,* p. 397.
52. Lakes painted a watercolor of Mudge chiseling out bones, YPM print no. 2085.
53. *Diplosaurus felix* is a junior synonym of *Goniopholis felix,* a crocodile, acc. no. 986 (catalogued YPM 517).
54. O. C. Marsh, "New Vertebrate Fossils," *American Journal of Science and Arts,* n.s., 14 (1877): 249–58; *Diplosaurus felix* is mentioned on p. 254.
55. Lakes to Marsh, June 27, 1877, *Marsh Papers,* reel 10, frames 416–26.
56. Lakes painted a watercolor of Mudge contemplating the huge bones while other members of the crew are depicted using picks nearby (YPM print no. 2084). See John Ostrom and John McIntosh, *Marsh's Dinosaurs* (New Haven, Conn.: Yale University Press, 1966), p. 25, fig. 6.
57. Lakes was mistaken in thinking that this specimen was the animal described by Marsh as *Apatosaurus grandis*. In the same paper in which Marsh founded his new genus and species, *Apatosaurus ajax,* from quarry no. 10 above, he also described a second species, *Apatosaurus grandis,* without stating its locality; "Notice of New Dinosaurian Reptiles from the Jurassic Formation," *American Journal of Science,* 3d ser., 14 (1877): 514–16. This second animal actually came from Como Bluff, Wyoming, in 1877. Marsh later referred *Apatosaurus grandis* to his genus, *Morosaurus,* now synonymized as *Cama-*

*rasaurus*. Lakes mistook *Apatosaurus grandis* to be the specimen YPM 4676 from quarry no. 8 at Morrison, now known to be a second specimen of *Apatosaurus ajax*.

58. This belongs to a second specimen of *Apatosaurus ajax* and is part of YPM 4676.
59. There are 48 species of *Gilia* listed as native to the Rocky Mountains but none with the name *vokus*. See Axel Ryberg, *Flora of the Rocky Mountains and Adjacent Plains* (New York: Hafner, 1954), pp. 690–92.
60. Thomas Sterry Hunt (1826–92) was a professor of chemistry and geology who then was teaching at the Massachusetts Institute of Technology. His research concerned the development of different minerals during the early history of the earth. See *Dictionary of Scientific Biography*. Leo Lesquereux (1806–89) studied fossil plants and worked for the USGS as well as a number state geological surveys. Lakes had been working as a collector of fossils for Lesquereux and the USGS and this request was more than a matter of courtesy. See letters from Lakes to Ferdinand V. Hayden, February 15 and July 12, 1877, from Hayden/Fryxell Collection no. 1638, box 53, American Heritage Center, University of Wyoming; *Marsh Papers,* June 20, 1877. reel 10, frame 408.
61. Professor Samuel H. Scudder (1837–1911) was the leading authority on *Orthoptera* (grasshoppers and related families). He published numerous works on entomology, which also included butterflies. During 1877 Scudder worked with the Hayden Survey in Colorado, Utah, and Wyoming collecting specimens. From 1886 to 1892, he was a paleontologist for the USGS, where he continued his work on fossil insects, many of which he collected from the Florissant Fossil Beds of Colorado. His collection from the Hayden Survey was eventually transferred to the United States National Museum (National Museum of Natural History, part of the Smithsonian Institution) during the 1890s. *Register of Samuel H. Scudder Papers,* record unit 7249, Smithsonian Institution Archives.
62. Bowditch is identified as F.C. in Samuel H. Scudder, *The Insects of North America* (Washington: Government Printing Office, 1890), p. 21. The Bowditch family was prominent in a number of fields of American science.
63. The Green River runs south from western Wyoming through northeastern Utah, where the White River joins it.
64. The museum (herbarium) was at the School of Mines. Lakes described it in his letter to Hayden, July 12, 1877: "The bulk of it was collected and presented by Rev. E. L. Green and contains the majority of the flora of these mountains. Additions have been also made by myself and Captain Berthoud." Hayden/Fryxell Collection no. 1638, box 53. Leaf miners live and feed in the mesophyll between the upper and lower surfaces of the leaf creating a galley or burrow visible beneath the epidermis of the plant.

65. These scientific names have been changed. Today locusts are regarded as either grasshoppers (order *Orthoptera*) or cicadas (order *Homoptera*).
66. At Cañon City, Oramel Lucas discovered a large fossil sauropod, *Camarasaurus supremus* (AMNH 5760), which he sold to Edward Drinker Cope. Cope hired Lucas to work for him. Mudge visited Lucas, took notes, and tried to persuade him to join with Marsh's crew. Mudge oversaw Marsh's diggings at a site not far distant on the property of M. P. Felch. From this Marsh quarry no. 1, Mudge sent specimens of *Diplodocus longus* and *Allosaurus fragilis*. In addition, it later produced specimens of *Ceratosaurus* and *Stegosaurus*.
67. This is an English butterfly, *Vanessa antiopa*. There are several members of the genus *Vanessa* that are native to Colorado: the red admiral *(V. atalanta)*; the western painted lady *(V. annabella)*; the cosmopolite *(V. cardui)*; and the American painted lady *(V. virginiensis)*.
68. Probably located near Conifer, Colorado, about twenty miles south of Morrison in Jefferson County. The Brooks are not mentioned in the 1880 Colorado census.
69. Elk Creek is the left tributary of the north fork of the South Platte River originating in the Park Range and flowing south east into the North Platte. See the *James Grafton Rodgers Collection*, Colorado Historical Society.
70. US Highway 285 passes through Kenosha Pass in Park County at an altitude of 10,000 feet.
71. Brubakers Station was on the route later taken by the Denver and South Park Railroad near Kenosha Pass. Bullwhacker was a nickname for teamsters. See the *Rodgers Collection*.
72. The use of a stagecoach was already anachronistic in many places in the eastern United States by the 1870s.
73. Probably Ferdinand V. Hayden's *Preliminary Field Report of the United States Geological Survey of Colorado and New Mexico* (Washington, D.C.: GPO, 1869). See in particular pp. 77–81.
74. Hamilton, founded in 1860 as a mining town on Tarryall Creek opposite the town of Tarryall at the junction of roads to mining camps to the north and South Park to the south was about fifteen miles northwest of Fairplay. By 1877 it had dwindled to a small mining camp. See Norma Flynn, *Early Mining Camps of South Park* (n.p.: n.p., 1953), pp. 5–6.
75. Fremont's Knoll or Summit in Park County is west of Michigan Creek southwest from Kenosha Pass and south and west of Jefferson. John Charles Frémont (1813–90) led his second exploration party back from Oregon and California through Hoosier Pass into the South Park in June 1844, during a period of intermittent warfare between the mountain Utes and the Arapahoes of the plains. They may have stopped here overnight but did not build fortifications. On the morning of June 23, Frémont and his men watched a

battle between Utes and Arapahoes. From his description of the defensive position, it appears to have been further south on the middle fork of the Platte River near Hartsel. See John Charles Frémont, *Report of the Exploring Expedition to the Rocky Mountains in the Year 1842 and to Oregon and North California in the Years 1843–44* (Washington, D.C.: Gales & Session, 1845), pp. 283–88; Ferol Egan, *Frémont: Explorer for a Restless Nation* (Garden City, N.Y.: Doubleday, 1977), pp. 253–56; Simmons, *Bayou Salado,* pp. 26–27, 58–60.

76. Possibly a grass spider of the genus *Agelenopsis,* which build such webs. The markings are difficult to identify with any genus.

77. Lakes refers in this journal to the Platte, which more precisely, is the South Platte River. Lechner's settlement, named for George Lechner, was a coal mine located about ten miles east of Fairplay; see the *Rodgers Collection*. Fairplay, the seat of Park County, was founded as a mining camp in 1861. By the late 1870s, it had become a center for stages, government, and commerce in the South Park; see Simmons, *Bayou Salado,* pp. 151–56.

78. Alma lies at the junction of Buckskin Creek with the middle fork of the South Platte River, about six miles from Fairplay in Park County. In 1877 it had a population of about seven hundred. See 1878 *Colorado Directory*.

79. Edward P. Arthur was an English-born, successful rancher in Park County. He had a son, born in 1880, named Peller. His ranch was located about eight miles southeast of Fairplay and just north of Garo. It later was known as either the Platte or Annex Ranch. Lakes gave Arthur's address for forwarding his mail. See Simmons, *Bayou Salado*, pp. 248–49; 1880 Colorado census; *Marsh Papers,* reel 10, frame 461.

80. John and Mary Williams are listed as having several children. He worked as a freighter and was of English descent. 1880 Colorado census.

81. Dudley was a mining encampment founded in 1872 about two miles north of Alma named for Judson Dudley, a developer of the Moose Mine on Mount Bross. See Flynn, *Early Mining Camps*.

82. See Ruth Ashton Nelson, *Handbook of Rocky Mountain Plants* (Tucson, Ariz: Dale Stuart King, 1969), pp. 26–36, for a discussion of plant succession and vegetation zones. This work is the basis for the citations listed here unless otherwise noted.

83. Possibly mountain parsley *(Pseudocymopteris montanus),* whiskbroom parsley *(Harbouria trachypleura),* or alpine parsley *(Oreoxis alpina)*.

84. Possibly whiteweed or whitetop *(Cardaria draba)*.

85. Probably mountain gooseberry *(Ribes inerme)*. The larvae were of the satyr angle wing or hop butterfly *(Polygonia satyrus)*.

86. Lakes tended to place butterflies on the right-hand side along with the altitudes in this listing, so this could be the parnassian butterfly *(Parnassius phoe-*

*bus)*. There are flowers called fringed parnassia *(Parnassia fimbriata)* or small-flowered parnassia *(Parnassia parviflora)*.

87. Elk thistle or drummond thistle *(Cirsium folosum)*.
88. Either Colorado columbine *(Aquilegia caerulea)* or alpine columbine *(Aquilegia saximontana)*.
89. Likely the big-rooted springbeauty *(Claytonia megarhiza)*, which grows in the alpine zone of the Rocky Mountains.
90. Sky pilot *(Polemonium viscosum)*.
91. Probably either the Rocky Mountain snow willow *(Salix reticulata)* or the rock willow *(Salix arctica)*.
92. There are several species of *Castilleja*, such as *Castilleja linariaefolia* (Wyoming paintbrush) and *Castilleja miniata* (scarlet paintbrush), which grow at lower levels, and *Castilleja rhexifolia* (subalpine or rosy paintbrush), which grows at higher levels.
93. Greasewood *(Sarcobatus vermiculatus)* usually occurs at lower latitudes and lower altitudes.
94. Possibly of the family *Clubionidae*, genus *Castianeira*. Several families of spiders, including *Clubionidae*, have members of the same species who often gather together in large groups at the beginning of cold weather and spend that period in closer company than usual for creatures noted for their cannibalism.
95. See note 85 regarding the larvae of *Polygonia satyrus*.
96. Dr. D. H. Dougan was a physician in Alma. The Russia Mine was located 500–800 feet south southeast from the summit of Mount Lincoln. It had pockety ore deposits caused by infiltration. Lakes noted in his survey journal in 1881 that the mine was shipping ten tons per month and was much worked with chambers the size of "mammoth caves." See Arthur Lakes, *Notes on the Mines and Geology of [the] Eastern Slope of Park Range from Mt. Lincoln to Buffalo Park*. United States Geological Survey, field notebooks, entry 90, record group 57, journal 1118, pp. 34–36, National Archives.
97. Frostwork are a type of anthodites usually composed of aragonite and have clusters of colorless to white needle or quill-like crystal spays that resemble spiny plants such as thistles. They grow on other cave formations. Lakes's article on them is unknown.
98. Added at a later date. See entries at end of this journal related to visit to Mt. Lincoln and the heading about the mountain rats, whose Colorado story was not recounted in these journals.
99. Quartzville was a mining encampment located on the east face of Mount Bross about ten miles from Fairplay along Colorado Highway 9 north of Alma. It flourished during the 1860s but was abandoned in the mid-1870s. See Flynn, *Early Mining Camps*.

100. Buckskin Creek originates from two small lakes, Kite-Shaped and Emma.
101. Located on the northwestern border of Park County at an elevation of 11,541 feet over the Park Mountains. Colorado Highway 9 travels this route north to Breckenridge in Gilpin County.
102. Lakes is referring to the Front Range of the Rocky Mountains, which rises abruptly from the plains.
103. At the bottom of this journal page that ends at this point, there is a marginal note: "mallard *Anas Boschus;* blue-winged teal, *Querquedula discors* [now known as *Anas discors*]; green-winged teal, *Querquedula carolinensis [Anas crecca];* red-breasted teal, *Querquedula cyanoptera [Anas cyanoptera].*"
104. Samuel Hartzel was one of the first ranchers in Park County, settling near the junction of the South and Middle forks of the Platte River about twenty miles southeast of Fairplay. He introduced cattle raising in the South Park during the 1860s and had numerous ancillary businesses, including a hotel with a hot spring resort. The Ute Indians regarded Hartzel as a particular friend and often could be found around the ranch still during the 1870s. The Hartzel ranch came to encompass more than 8,700 acres by 1881. See Simmons, *Bayou Salado,* pp. 30, 216–22.
105. These markers were used to establish survey points. A survey party of Hayden's had been in South Park in 1873 and Hayden used Fairplay as a rendezvous for survey parties that year. See Richard Bartlett, *Great Surveys of the American West* (Norman: University of Oklahoma, 1962), pp. 109, 310.
106. Jerome Harrington moved to South Park from New York with his family in the early 1870s and established a ranch that came to consist of more than 6,000 acres, located about five miles from Hartzel's ranch. See Simmons, *Bayou Salado,* p. 241.
107. Lakes is referring to *Cyrene Lamarck* 1818, *Corbicula Megerle* 1811, and *Mactra Linnaeus. Mactra canonensis* is not part of the genus *Mactra* and may be a superseded species.
108. Sulphur Springs in Park County was located southwest of Sulphur Mountain about eight miles from Hartsel. It was also known as Eagle Sulphur Springs and later Oliver Hot Springs. Efforts to develop a spa there failed during the 1880s. See Simmons, *Bayou Salado,* p. 241.
109. Hydrogen sulfide, which smells like rotten eggs, is a highly toxic gas.
110. The Tertiary began 65 million years ago and continued until 1.6 million years ago. For a description of the geology of much of the South Park, see J. T. Stark et al., *Geology and Origin of South Park, Colorado* (Baltimore: Geological Society of America, 1949).
111. Another species of the Cretaceous clams, noted for concentric ridges on their shells.
112. Judge James Castello settled in South Park during the early 1860s. He served as county judge, member of the first Colorado Territorial Senate (1865) and

receiver for the United States Land Office c. 1865–72. He settled a community named after his city of origin, Florissant, Missouri, located about fifty miles from Fairplay. See p. 000.

113. In 1874 Adam G. Hill, and his wife, Charlotte, bought the homestead of Rev. David P. Long from Long's son-in-law, Frank Sens. It was located about a mile and half south of Castello's ranch at Florrisant. Rev. Long had homesteaded there beginning in 1871 or 1872 but then moved on to Salt Lake City. See Jim McChristal, *A History of Florissant Fossil Beds National Monument: In Celebration of Preservation* (Denver[?]: National Park Service, 1994), pp. 3–4; Colorado 1880 census.
114. This site is now known as Fortification Hill in Florissant. Although no specific event can be associated with this story, Judge Castello ran a general store and trading post that dealt with the Utes from whom he likely heard this story. Conversation with Ms. Doris Kneuer, Volunteer Curator, Florissant Fossil Beds National Monument, October 13, 1995. For a general discussion of the Ute-Arapaho conflict, see J. Donald Hughes, *American Indians in Colorado,* 2d. ed. (Boulder, Colo.: Pruett Publishing, 1987), pp. 26–37, 74–75.
115. According to the 1880 census, Jacob Houghton was working as a mining superintendent.
116. Amazonite or amazon stones are colored a variety of shades of green. Its structure is monoclinic: a potassium rich feldspar. These stones are found in the Pikes Peak region.
117. This site may be misnamed. The Great Pyramid is a strictly geometric pyramid as compared with the step pyramids of earlier pharaohs such as Zoser. See Ahmed Fakhry, *The Pyramids* (Chicago: University of Chicago, 1969).
118. George Montague Wheeler (1842–1905) was in charge of the United States Army's United States Geographical Surveys West of the One Hundredth Meridian, 1872–78. One of his survey parties, under the command of Lieutenant William Marshall, surveyed part of South Park during the summer of 1873 at the same time and where one of Hayden's survey crews under the leadership of Henry Gannett was working. This incident was one reason for the investigation of the surveys by Congress in 1874. See Bartlett, *Great Surveys,* pp. 334,368.
119. Lakes was mistaken regarding his geography. Bergen Park is north of South Park near Idaho Springs. Hayden Park is in the southwestern park area of Teller County, west of Pikes Peak, and is also known as the Florissant Valley.
120. Among the trees fossilized were giant sequoias, with in one instance a trunk 22 feet in diameter. There was a great deal of undocumented removal of fossil specimens from Florissant with material going to exhibitions and museums as well as to decorate homes and businesses. Arthur Lakes commented upon the extent of the destruction during the twenty years since his first visit to South Park in "The Florissant Basin," *Mines and Minerals* 20 (1899): 179–80.

The Florissant Fossil Beds National Monument was established in 1969 to manage this site. See McChristal, *History of Florissant Fossil Beds,* pp. 3–34.

121. Any of various chiefly tropical insects of the subfamily *Fulgorinae* having an enlarged elongated head.
122. Of Frémont's five expeditions, four crossed parts of Colorado. His easy manner with frontiersmen and love of the West created good will for him in that region. See Egan, *Frémont*. Lakes was mistaken about Costello's heritage in that he was Irish and Spanish rather than French. See *Florissant Colorado,* p. 1.
123. Metallic wood-boring beetles of the family *Buprestidae*.
124. A. C. Peal, "Geology of South Park," in *United States Geological and Geographical Survey of the Territories. Annual Report. 1874* (Washington, D.C.: GPO, 1874), pp. 193–239.
125. Twin Creek runs westward from the Front Range just north of Florissant.
126. Known as the Royal Gorge.
127. US Highway 24 now passes over it in Teller County at elevation 9,165 feet.
128. The Silurian Period was 439–409 million years ago; the Lower Silurian was 439–430 million years ago.
129. "August 24" is written here and "Left M" crossed through.
130. Manitou Springs (elevation 6,250 feet) had a population of about two hundred in 1877 and was noted for its mineral springs. Edward Erastus Nichols was the proprietor of the Cliff House Hotel, which received its name from the cliff dwelling ruins found in the vicinity that were built during the Great Pueblo Period, 1100–1300 A.D. See Bettie Marie Daniels and Virginia McConnell, *The Springs of Manitou* (Manitou Springs, Colo.:, Manitou Springs Historical Society, 1982), pp. 1–20; 1878 *Colorado Directory*.
131. The ground hog or woodchuck *(Marmota monax)* is an eastern woodlands creature. The yellow-bellied marmot of the western mountain regions has as its current scientific name *Marmota flaviventris*.
132. The weather station was occupied by the US Signal Corps from 1873 to 1889. There were accommodations for tourists only after the departure of the Army. Sergeants C. H. Hobbs and John O'Keefe were stationed there during this period. See Daniels and McConnell, *Springs of Manitou,* pp. 41–45.
133. S. Edwin Solly is listed as practicing medicine in Colorado Springs opposite the Bacons' Hotel.
134. The Garden of the Gods and Glen Eyrie rock formations near Colorado Springs have exposed strata ranging from the Pennsylvanian (280–230 million years ago) to the Cretaceous (135–63 million years ago). Its unique geological feature is the spectacular faulting that has created this scenic topography. See Jeffrey Noblett, *A Guide to the Geological History of the Pikes Peak Region* (n.p. [Colorado College]: 1994). The Crawford House had D. C. Crawford as its proprietor. *Colorado Springs Gazette*, February 12, 1876, p. 1.

135. Rev. J. F. Walker was the Episcopal minister in Colorado Springs. Dr. Thomas Huxley was a leading proponent and defender of Darwin's theories.
136. Thomas Sterry Hunt (1826–92) had written to Lakes regarding the visit of Scudder and Bowditch. His research concerning the development of different minerals during the early history of the earth focused upon the action of water, sedimentation, and chemical reactions. See *Dictionary of Scientific Biography*.
137. The Archaean (Archean) are the oldest rocks (approximately 380 million years) of the Precambrian period. Laurentian rocks are gneissic granite of the early Precambrian. Hunt was correct in this statement.
138. F. W. Beebee ran a hotel built in 1875 first named the Mansions and then the Beebee House, located just west of the present Town Hall. Hanson Risley is listed as living in Colorado Springs. 1880 Colorado census. See Daniels and McConnell, *Springs of Manitou,* pp. 23–25.
139. The American Association for the Advancement of Science met in Nashville in late August 1877.
140. The periods identified as Tertiary 1 and 2 were considered to be Eocene, the Paleocene period then not having been recognized. In the Rocky Mountains they were called the Wasatch and Bridger Group by Hayden. The Monument Creek Group is an unidentified group but may be analogous to the Florissant Group since Lakes placed it above the Bridger Group, where the Florissant Group, also known as the Uinta Group, was located. James D. Dana, *Manual of Geology* (New York: American Book Company, 1895), pp. 884–93.
141. The Monument House Hotel, whose proprietor was Dr. Robertson, was about twenty miles north of Colorado Springs. Divide is fifteen miles from Florissant, located just west of Ute Pass. It is not near the continental divide, which is further west of the Front Range. Lakes reference is to the top of the Front Range.
142. An allusion to Alfred, Lord Tennyson's "Charge of the Light Brigade."
143. Greenland was about five miles from Divide on the Denver & Rio Grande Railroad.
144. There are no buzzards in the New World. Lakes is referring to the northern harrier *(Circus cyaneus hudsonius),* also known as the American harrier or marsh hawk.
145. The phrase "(Mr. Allen's dog)" is added to the text at this point in Lakes's handwriting, possibly to remind him to recount an anecdote concerning the dog. There is no story about Allen's dog.
146. There was a good deal of lawlessness related to the gold strike in the Black Hills during the 1870s with "road agents" robbing the stage coaches that hauled all the supplies into and gold out of the Dakota Territory.
147. Thomas Harris is listed as proprietor of the Castle Rock Hotel according to

the 1877 *Colorado Directory*. He and his wife, Mary, both immigrated from England according to the 1880 census, which listed Judge John H. Craig and Charles Holbrook, a lawyer, as boarders of the hotel. Anna and Hiram Foster are listed as keeping the Foster House, whose boarders the census identifies as laborers. See 1880 Colorado census.

148. Now known as Devils Head, altitude 9,348 feet, located in Douglas County. See the *Rogers Collection*.
149. Castle Rock is in Douglas County along the Denver and Rio Grande Railroad, thirty-three miles south of the Denver station and forty-two miles north of the Colorado Springs station. The 1877 *Colorado Directory* listed a courthouse and a Methodist church.
150. Mt. Evans, altitude 13,557 feet, is part of the Mosquito Range in Clear Creek County, two miles south of the ridge of Mosquito Pass.
151. Littleton was then ten miles south of Denver station on the Denver and Rio Grande Railroad.
152. Quotation from Tennyson's "Northern Farmer Old Style" (1864): "But summun 'ull come ater meå mayhap / wi' 'is kittle o' steåm / Huzzin' an' maåzin' the blessed feåld wi' / the divil's oån teåm" (stanza 16).
153. Deer Creek is about six miles south of Morrison, Colorado.
154. Professor Cyrus Fogg Brackett (1833–1915) served as the Henry Chair of Physics and chair of the physics department at Princeton University from 1873 to 1908. Brackett was also a physician and served as president of the New Jersey State Board of Health from 1888 to 1908. He organized the School of Electrical Engineering at Princeton. Communication from Susan Illis, Princeton University Archives, December 13, 1995.
155. The word "anxio" is crossed through here. William E. Nugent from Morrison is listed as age 23 and working in the lime business in the 1880 Colorado census. Nugent worked for Marsh at this quarry through the spring of 1879.
156. Mudge sent thirteen boxes (acc. no. 1009) from quarry no. 5, consisting of remains from *Stegosaurus armatus* (YPM 1850) as well as unidentified sauropod material. An additional seven boxes of similar material were sent on October 11, 1877. Most of this material has not been prepared as of 1996.
157. Major General Benjamin Franklin Butler (1818–93) of the Union army incurred the bitter censure by southerners during his occupation of New Orleans in 1862. Besides reestablishing federal authority, his troops are reputed to have stolen personal property from southern households. He is most noted for his order (no. 28) that southern women who insulted his troops should be treated as women of the town. Jefferson Davis outlawed him.
158. J. G. Pease was the proprietor of the Morrison House Hotel. He was the Democratic party candidate for sheriff in Jefferson County. The fall 1877 election in Colorado included a referendum on whether to extend the vote to women, which probably was the reason the women visited these potential

electors. The vote was not extended. See *Colorado Transcript,* October 3 and 10, 1877.

159. Lakes conducted itinerant services in Alma, Leadville, Idaho Springs, and Fairplay. Breck, *Episcopal Church in Colorado,* p. 82.
160. Probably *Pica pica.*
161. Possibly the red-naped sapsucker *(Sphyrapicus nuchalis).*
162. Quarry no. 10.
163. These were part of *Apatosaurus ajax* (YPM 1860). The sacrum is plate 29 in Ostrom and McIntosh, *Marsh's Dinosaurs.* The scapula was actually a pubis.
164. This femur bone came to be a source of considerable embarrassment for Marsh. After Lakes reported that he had found the other end of the femur, Marsh made inaccurate completed bone models of *Atlantosaurus immanis* that were each about eight feet long. They turned out to be too long when compared with newly discovered intact femur bones of the "Brontosaurus" = *Apatosaurus,* and Marsh was required to exchange the models with scaled-back versions that were still about six feet long. It is correctly figured in O. C. Marsh, "Dinosaurs in North America," *United States Geological Survey. Sixteenth Annual Report of the Secretary of the Interior. 1894–5,* p. 166, pl. 16, figs. 2 and 3; Charles Walcott, Part I. Director's Report and Papers of a Theoretic Nature.
165. Samuel Wendell Williston (1851–1918) received a B.A. from the Kansas State Agricultural College in Manhattan (Kansas State University). He had been a Mudge's student and was hired by Marsh to help him collect fossils in Kansas beginning in 1876. He worked for Marsh until 1890. Williston graduated from Yale in 1880 with a degree in medicine. He taught at the University of Kansas from 1890 to 1902, and then at the University of Chicago from 1902 to 1918. Williston made significant contributions both in paleontology and entomology, and he published research on his own time while working for Marsh. See Elizabeth Noble Shor, *Fossils and Flies: The Life of a Compleat Scientist Samuel Wendell Williston* (Norman: University of Oklahoma Press, 1971), pp. 87–88, for a description of Williston's activities in Colorado in 1877.
166. In December 1877 Marsh sent Lakes a telegram telling him to finish up work by the end of the month. See *Marsh Papers,* reel 10, frame 491.
167. The Evergreen Hotel and the Morrison House Hotel were in Morrison.
168. Jarvis Hall or the Colorado School of Mines.
169. Probably a passenger train from Omaha, which would have taken the route of the Union Pacific to Cheyenne, then the Colorado Central Main Line to Denver, and then either its Mountain Division to Golden or the Denver, South Park and Pacific to Morrison.
170. Lakes's estimation of the size of the *Apatosaurus immanis* as being seventy feet long was extremely perceptive. Its immense size was generally not recognized at this time.

171. Edwin Drinker Cope, *Report of the United States Geological and Geographical Survey of the Territorries: The Vertebrata of the Cretaceous Formations of the West*, vol. 2 (Washington, D.C.: GPO, 1875). Here begins a series of sections that Lakes wrote probably some time after the field journals. Lakes summarizes Cope's report.
172. Ibid., pp. 22–23, 43. *Haploscapla* is a junior synonym of *Inoceramus*.
173. The description of vertebrates covers pp. 53ff. of Cope's report
174. This is a description of a mosasaur.
175. Ibid., pp. 46–48.
176. Genus of mosasaur; ibid., pp. 142–44.
177. Genus of *mosasaurus;* ibid., pp. 48, 130–39.
178. Another species of *Clidastes*. Ibid.
179. Ibid., p.48.
180. Ibid., p. 48–50
181. The genus is now known as *Xiphactinus*.
182. Cope's theory of the extinction of marine reptiles reflects commonly accepted opinion of many late nineteenth-century scientists who believed that the evolution and extinction of species was gradual. Its manner and degree are still fiercely debated.
183. Cope, *Report,* pp. 7–10, part 1 of the introduction. The contents are dated by over a century of scientific research. No effort has been made to comment upon every questionable theory or assumption, but the reader should be aware that much of it is outdated.
184. The living members of the order *Artiodactyla* (even-toed ungulates) can be divided between the *Ruminanta,* with molars assuming the shape of a crescent (selenodont condition), two toes, and a compound stomach, and *Suina,* with molars having a gently rounded cusps (bunodont condition), four toes, and a simple stomach. See Alfred Sherwood Romer, *Vertebrate Paleontology* (Chicago: University of Chicago, 1945), pp. 442–46; Robert Carroll, *Vertebrate Paleontology and Evolution* (New York: Freeman, 1988), p. 510.
185. Cope, *Report,* pp. 11–14, part 2 of introduction. Again, the information is dated.
186. The ancestral genus for the following animals appeared during the following periods: elephant in Pliocene, rhinoceros in Miocene, opossum in Pleistocene, raccoon in Upper Miocene, cat in Upper Miocene, wolf in Lower Pliocene, fox in Upper Miocene/Lower Pliocene, and weasel in Upper Miocene.
187. Cope, *Report,* p. 15.
188. Ibid., p. 16.
189. After the heading "Getting a *Cinclus* Nest in Bear Creek Canon," there are twenty-one blank pages before the text resumes with recollections based upon events during 1878. Lakes mistakenly wrote "1868" in the date for the

trip to South Park, which should be April 7, 1878. During that spring, Lakes was still working for Marsh at Morrison, teaching at Jarvis Hall, and ministering to various mining communities.

190. Possibly Aison Allen, who is listed as a rancher in South Park. See 1880 Colorado census,
191. The Old World swallowtail *(Papilio machon)* is not found in the Americas; these were likely *Papilio polyxenes*.
192. The *Denver Daily Times,* for example, had a short article about the fire on April 4, 1877. Jarvis Hall was destroyed by fire on April 4, which started in the attic. Two days later a fire that started in the belfry consumed Matthews Hall. Newspaper accounts (*Colorado Transcript,* April 5 and 8, 1878) indicate that the Matthews Hall fire at least was blamed on arson. The result was that the insurance money was used to transfer and rebuild Jarvis Hall in Denver. See Breck, *Episcopal Church in Colorado*.
193. Mount Silverheels is about fourteen miles from Kenosha Pass.
194. The settlement, named for George Lechner, was a coal mine located about ten miles east of Fairplay. See the *Rodgers Collection*.
195. Easter was on April 21 in 1878. Aison and Charlotte Allen had several children, including one who is listed as four years old in the 1880 Colorado census.
196. Probably the spotted sandpiper *(Actitis macularia)*.
197. Possibly the Mr. Munrow that Lakes mentioned earlier. The Methodist church in Alma had Rev. J. L. Dyer as its minister.
198. Dr. Dougan was the owner of the mine. Lakes climbed Mt. Lincoln with Scudder the August before. Mudge visited Lakes during the spring of 1878.
199. Lakes apparently wrote a scientific paper concerning the mountain rats. He submitted it to Marsh in the early 1880s accompanied by a sketch of the rats' nest and a "fancy sketch of them carrying away our hammers by moonlight." He inquired about the status of the article in a letter to Marsh dated March 21, 1885. The article and sketches were never published and are lost. It is possible that Lakes stopped writing recollections in his journal in order to prepare this article. See Incoming Correspondence, Assistant Secretary in charge of the United States National Museum, 1860–1908, record unit 189, item 24648, box 71, Smithsonian Institution Archives.
200. There are several loose small pages in the back of the journal: pl. 5 from the *American Journal of Science* 8 (1879) of *Morosaurus grandis;* two unidentified rough sketches of bones; sketches possibly of prairie dogs; a small piece of paper with the words "National Museum For Work" and "Scelidosaurus" written on it. *Scelidosaurus* was a primitive *Stegosauriod* found in Lower Jurassic. Finally, there is the following note: "self importance and the extreme peril of himself and the community he presents. His watchtower is usually an angular block of grey lichened trachytic porphyry with which the

summits of these Silurian mountains forms the South Park range are usually capped broken up by frost and other denuding agencies into piles of angular fragments like fallen ruins.

"The home of the coney is a region of eruption and erosion of eruption which has folded these mountains and crumpled them into long billowing folds called ranges the lava oozing out in the process and erosion which has sculptured them into all the glories of the mountain landscape. Great patches of snow rim the edges and slopes of wide and deep amphitheaters, great *cul de sacs* and Devil's punch bowls scoop out near every mountain top homes of ancient glaciers and theatres of an erosion whose work in these Rocky Mountains is almost incredible. Cubic miles of rock have been removed and [in] the hills the coney lives."

## Journal of Explorations for Saurians and Fossil Remains in Wyoming, 1879–1880

1. In late April 1879, Lakes received a telegram from Marsh directing him to discontinue operations at Morrison and travel to Wyoming. *Marsh Papers,* reel 10, frame 533.
2. Sometime after writing his journal, in the early 1880s, Lakes copied into it portions of letters that he had written to his family in England. Only portions relating to his activities in Wyoming exist and there is no mention of family matters. The letters date from May 23 to June 24, 1879, and are on pp. 101–30 of the Wyoming journal. Where the version of events given in the English letters adds substantially to the narrative, it here replaces the corresponding journal version, which then appears in the notes.

   In this instance, two paragraphs from a letter of May 23, 1879, replace the original journal version, which reads as follows: "I left Golden City by train for Cheyenne. After passing in the distance the grand display of Triassic red rocks forming the cathedral spires of Boulder, the basalt dyke of Valmont and along the railroad a number of baby towns just springing into existence and also our old fossil ground on Fossil Creek near Fort Collins, the rest of our way to Cheyenne was over a monotonous rolling prairie." Valmont was a former railway station on the Union Pacific Railroad in Boulder County, Colorado, about four miles east the city of Boulder with a Valmont Butte nearby. Fossil Creek is located about three miles south of Fort Collins and contains fossiliferous Morrison Formation strata.
3. Fingal's Cave is located in the Inner Hebrides on the island of Staffa, about seven miles west of the island of Mull. The cave is about 227 feet deep, 42 feet wide, with the sea covering the floor to a depth of 25 feet at low tide. It has hexagonal basaltic columns supporting its roof which is 66 feet above sea level at low tide.
4. *Inocerami* were pelecypods common in the Jurassic and Cretaceous oval in

outline with concentric, corrugated, growth circles; ammonite is a common name for a large number of Mesozoic cephalopods that often had spiral shells. Scaphiles are Cretaceous cephalopods with a flattened spiral coil. Frank Rhodes, Herbert Zim, and Paul Shaffer, *Fossils: A Guide to Prehistoric Life* (New York: Golden Press, 1962), pp. 120, 124–30.

5. Fort Sanders was located by the Union Pacific Railroad next to Laramie City.
6. Probably the Wind River Reservation of the Shoshones in west central Wyoming.
7. Reed sent seven shipments to Yale, comprising some 150 specimens, between February 15, 1878, and May 7, 1879.
8. *Camarasaurus grandis* (YPM 1907).
9. *Allosaurus lucaris* Marsh. It later became the type specimen for the genus of carnosaurs *Labrosaurus*.
10. The Ichthyosaurus genus *Baptanodon* replaced *Sauranodon*, which had been used to name another creature.
11. The *Atlantosaurus* horizon is now known as the Morrison formation, which has produced numerous Upper Jurassic fossils. The Upper Jurassic was approximately 157–146 million years ago.
12. The Lias formation is from the Lower (Early) Jurassic, approximately 208–178 million years ago. In Europe numerous fossils of *Ichthyosauria* exist from this period. Actually the *Sauranodon* beds, now known as the Sundance Formation, are higher than the Lias and are Middle Jurassic (178–157 million years ago). Lakes was mistaken in equating them specifically with the Lias. Belemnites are cephalopods from the Mississippian to Cretaceous periods that are in the form of cigar or bullet-shaped fossils. Rhodes, Zim, and Shaffer, *Fossils,* p. 130.
13. This is a common fossil phenomenon resulting from wave action.
14. Here a paragraph from Lakes's letter to England dated May 16, 1879, replaces the original journal version, which reads: "As we walked home, Reed showed me a trestle bridge over a ravine some twenty feet deep where a gang of desperadoes the preceding winter had tried to wreck a train by pulling out the rails. Fortunately, their scheme was detected before the train arrived. He also showed me a ravine called Robber's Roost where the same gang camped to prepare for their wrecking scheme. A sheriff and another man who undertook to follow the desperadoes were both shot dead from an ambuscade."
15. Here a paragraph from an English letter of May 18, 1879, is substituted for the original journal version, which reads: "Our tent was struck and our things speedily packed. The party moved off over the hill, one carrying our sheet-iron campstove on his back, another our tent poles and a third our kitchen utensils, etc. It was a somewhat comical sight to see the party winding over the bluffs like a tribe of white Indians carrying their household goods. The tent was soon pitched near the section house not far from the

shore of the lake and for the rest of the day it was crowded with railroad hands stretched out smoking their pipes, chatting and laughing; a reinforcement from a station below swelling their numbers till the tent could not hold them all.

"They appeared to be rough, good natured fellows, not very choice in their language. The profanity of men in this wild region strikes a stranger at first as something frightful till after a time one becomes in a sense accustomed to it. Reed and I dined with the men at the section house to avoid the trouble of cooking a meal for ourselves."

16. This sketchbook was accessioned by the Smithsonian Library in 1900 along with Lakes's journals, but it is now missing.
17. Edward Kennedy (1853—?) worked for the Union Pacific and unofficially helped Reed and Marsh's party during the summer of 1879. In 1880 Marsh hired him and Kennedy worked for Marsh until 1884. Ostrom and McIntosh, *Marsh's Dinosaurs,* pp. 21–22, 27–29, 35–38, 41–46; 1880 Wyoming census.
18. *Ambystoma tigrinum* or Tiger Salamander, also known as *Ambystoma mexicanum* and *Siredon mexicanum*. A variant in the spelling during the 1880s was *Amblystoma mexicanus*. North American salamanders of the genus *Ambystoma* unlike most amphibians, often retain their external gills and become sexually mature without undergoing metamorphosis, known as neoteny. The true Mexican axolotl is only found in Central Mexico. J. F. D. Frazer, *Amphibians* (London: Wykeham Publications, 1973), pp. 54–55; Roger Conant, *A Field Guide to Reptiles and Amphibians of Eastern and Central North America* (Boston: Houghten Mifflin, 1975), p. 256.
19. This was later known as quarry 1½. The sacrum is that of *Allosaurus fragilis* (YPM 4840).
20. William Edward Carlin worked as station master at Como before working for Marsh 1877–78, collecting fossils along with Reed. By 1879 Carlin was working for Cope and on bad terms with Reed. Ostrom and McIntosh, *Marsh's Dinosaurs,* pp. 6–24. Lakes consistently spells his name "Carline."
21. There were seven Yale Sauranodon quarries.
22. Here three paragraphs from an English letter of May 20, 1879, replaces the original journal version, which reads: "Reed and I went with the men on the handcar up the railway track for some distance towards No. 3 Quarry and we spent the day in further uncovering the skeleton. Most of the bones were very rotten and the work progressed slowly.

"The bones appeared to be mostly vertebrae and ribs: three cervical, a dorsal and several caudal vertebrae. They were all jet black in color.

"On our way home, we revisited the site of our first camp. Reed showed me the skeleton of an elk he had killed. As we walked through the sagebrush on a little hillock in front of it, the long necks of a dozen sage hens [sage grouse *(Centrocercus urophasianus)*] popped up staring foolishly around them. Reed shot one with his rifle. It was a handsome bird about the size of a large pheasant. Its fan-like tail runs out into sharp points. The plumage is a grey granite color. A white ring around the neck showed it to be a male bird."

23. The bones that Lakes mentions were from either an *Allosaurus lucaris* or *Morosauris grandis,* now known as *Camarasaurus grandis*.
24. Here two paragraphs from an English letter of May 21, 1879, replace the original journal version, which reads: "Started early on the handcar to No. 3 and spent the day in further uncovering the saurian and in sketching the skeleton as it lay with Reed getting along side of it.

    "At supper we had canned strawberries and shortcake as a great treat. The railroad hands came in after supper and I amused them with my sketchbook and sketched one of them named Kennedy as a figure in my picture of Saurian Quarry No. 3. "We had an exhilarating ride on the handcar to our quarry in the early morning. Five strong fellows pumping at the handles made the car fly over the track. Numbers of spermophiles and rabbits startled by the noise fled for refuge among the sagebrush. We uncovered several large bones, amongst them two which appeared to be an ischium and an ilium [in fact, a scapula] lying close together; also some vertebrae and large ribs. At noon we cooked our coffee and studied one of Marsh's pamphlets on dinosaurs which he had sent me. Reed and I also had a talk on the subject of evolution."
25. Possibly watercolor of Reed and Kennedy excavating bones, YPM print no. 2079; Ostrom and McIntosh, *Marsh's Dinosaurs,* p. 27, fig. 9.
26. George, not John, Leech, age 56, worked as a snowshed tender for the Union Pacific at Rock Creek. See 1880 Wyoming census.
27. David C. Chase, age 27, was listed as the railroad telegraph operator in the 1880 Wyoming census, although Lakes consistently refers to him as the station master.
28. Here six paragraphs from an English letter of May 23, 1879, replace two paragraphs in the original journal version, which read:"Blowing hard: heavy work for the men on the handcar beating up against the wind. Packed up the bones in gunny sacks and carried them to the railroad track. We flew home to camp on the handcar with the wind at our back.

    "The afternoon was spent in making and packing a box with our specimens and dispatching them to Yale [YPM acc. no. 1231]. Reed doffed his buckskin coat and emerged from the tent in a most respectable dress as he was going to the little town of Medicine Bow for supplies."
29. David C. and Mary Chase.
30. F. F. Hubbel and his unnamed brother from Michigan worked for Cope at Como Bluff in 1879–80. See Ostrom and McIntosh, *Marsh's Dinosaurs,* pp. 24, 36; Henry Fairfield Osborn, *Cope: Master Naturalist* (New York: Arno Press, 1978), p. 259.

31. Edward Ryan, age 39, and his wife, Johanne, age 37, are listed in the 1880 Wyoming census as having Ireland as their birthplace. His job is recorded as railroad section foreman.
32. There was a major gold strike in the Black Hills of Dakota Territory in the late 1870s and a similar silver strike in the Leadville area, in the Mosquito Range of Colorado beginning in 1877. In the aftermath of these rushes, large numbers of fortune seekers were disappointed and penniless.
33. Probably the spotted sandpiper *(Actitis macularia)*.
34. The heath family's name is Ericaceae.
35. Possibly the Watson pennstemon *(Penstemons watsoni)*.
36. *Camarasaurus grandis* (YPM 1907).
37. A basin area between the Medicine Bow Mountains and the Rocky Mountains in Jackson County in north central Colorado.
38. See "Skeleton Found in Sandstone," *Daily Boomerang* (Laramie), October 12, 1911, for an account of the discovery of a similar skeleton.
39. Left humerus of *Camarasaurus grandis* (YPM 5858). Here begin two paragraphs from an English letter of May 29, 1879, which replace a phrase and sentence in the original journal version: "also the core of the claw of a carnivorous dinosaur from six to eight inches long. I made a sketch of the claw core and also found a tiny limb bone two and half inches long."
40. *Allosaurus lucaris* (YPM 1931).
41. Lakes is referring to the British weight one stone, which equals 14 pounds.
42. This is a typical crocodile fossil, probably *Goniopholis*.
43. Lakes's Sunday reading was Robert Chambers, *Vestiges of the National History of Creation* (London: Churchill, 1844).
44. Edward G. Ashley (1845– ?) worked for Marsh 1879–82. See Ostrom and McIntosh, *Marsh's Dinosaurs;* 1880 Wyoming census.
45. These turned out to be a crocodile *Goniopholis*.
46. The two vertebrae six inches long are two sacrals of *Camarasaurus grandis* (uncatalogued). The identity of the small vertebrae is uncertain but they are probably distal caudals of *Camarasaurus*.
47. An observation car, from "rubberneck car" for passengers looking around or "rubbernecking."
48. *Allosaurus lucaris* (YPM 1931).
49. Only crocodile bones of *Goniopholis* are available from quarry no. 6. The small bones of the possible pterodactyl or mammal may be lost or uncatalogued.
50. Garden snails are of genus *Helix,* family *Helicidae.*
51. A sauropod now known as *Camarasarus grandis*. As of 1996, the specimen from quarry no. 4 had not been prepared.
52. They were the type specimen YPM 1901 and paratype YPM 1905 of *Camarasaurus grandis*.

53. These have not been identified among the Yale collections.
54. *Atlantosaurus immanus* = *Apatosaurus ajax*. The cavities are a characteristic weight reduction feature of a number of large dinosaurs.
55. In 1870 Marsh first took a group of Yale students out to Fort Bridger in southwestern Wyoming and to other western fossil localities. He subsequently led expeditions to this area, rich in Eocene fossils, from 1871 to 1873. For a popular account of these expeditions, see Robert Platte, *The Dinosaur Hunters* (New York: McKay, 1964), pp. 99–138.
56. Family of theropods from the Jurassic, including the genus *Allosaurus*. This quarry produced a specimen of *Allosaurus fragilis* (YPM 4840).
57. Watercolor of Ashley, Reed, and Marsh, YPM print no. 2086; Ostrom and McIntosh, *Marsh's Dinosaurs,* p. 25, fig. 7.
58. The exact location of sauronodon quarry 1 is unknown other than it is near Robber's Roost.
59. The site of the original *Dryolestes priscus* (YPM 11820) is not recorded. It is possibly quarry no. 7 but more likely a more separate site where nothing else was found.
60. This became known as quarry no. 7 or the Three Trees Quarry. The Upper Jurassic ornithropod dinosaur discovered here was known as *Laosaurus consors* (YPM 1882) and has now been synonymized as *Dryosaurus altus* by Peter Galton. See Peter Galton, "*Dryosaurus,* a Hypsilophodontid Dinosaur from the Upper Jurassic of North America and Africa," *Palaeontologische Zeitschrift* 55 (1981): 271–312.
61. For contemporary accounts, see "The North Park Find," *Cheyenne Daily Sun,* June 13, 1879, p. 4; "North Park Mines: Return of Dr. William Davis and J. Kaufman—What They Saw and Heard," *Cheyenne Daily Sun,* June 27, 1879, p. 4.
62. Here four paragraphs from an English letter of June 9, 1879, replaces the following original journal text: "Reed's birthday. We went to Three Trees Quarry and found several bones amongst them some toe bones consecutive and united; also a portion of a leg."
63. Here two paragraphs from an English letter of June 11, 1879, replaces the following original text: "Ashley and I went to No. 6 and prospected along the bluff towards the west finding a few bones. A heavy thunderstorm came on. We took refuge under some rocks at the top of the bluff. It was a grand sight. A sharp peal followed by a flash of forked lightning seemed to strike the bluff just behind us. From our shelter we watched the storm passing off towards the North. Heavy rain fell followed by hail, the stones as large as pigeons eggs. As the storm passed off, a few joyous notes of birds rang out as if rejoicing that the storm was over. All nature seemed refreshed."
64. Here four paragraphs from an English letter of June 12, 1879, replace the following original text from the journal: "Cold and drizzling. Prospected for

bones finding two large carnivorous teeth and also a long herbivorous one. We dined at No. 6. Whilst we were eating our dinner, we spied a coyote skulking towards the bluff. Reed was soon in chase and killed him. The hide is not valuable at this season.

"At No. 6 we found some small vertebrae and other bones all hollow which we thought might be those of a pterodactyl. Heavy thunderstorm. We walked home through streams of water and reached camp well drenched. The water from the bluffs was white as milk."

65. Although the morosaurus = camarasaurus was a herbivore like iguanodon, its teeth were not similar to the latter creature. The teeth of the carnivorous dinosaurs megalosaurus and allosaurus are somewhat similar.
66. Only crocodile fossils *Goniopholis* have been identified from quarry no. 6.
67. These mammalian fossils have not been identified and there is no record of them.
68. Here three paragraphs from an English letter of June 13, 1879, are included to supplement the journal account of the same day. One paragraph of the original journal text has been removed: "George Leech, an Englishman who keeps the snowshed about six miles north of Como, came and chatted with us at the quarry. He lives a lonely batchelor life, his only companion is his dog. Consequently he dearly likes an opportunity for a chat. Reed spent the afternoon loading ammunition."
69. The humerus YPM 5858 and scapula YPM 6217 are of *Camarasaurus grandis* = *Morosaurus grandis* as are the claws. It is a small specimen of this creature. The vertebrae are of *Allosaurus lucaris.*
70. Here two paragraphs from an English letter of June 14, 1879, replace three sentences of text fom the original journal: "Went to No. 3 and packed up bones. We found a number of consecutive caudal vertebrae below the scapula, also a carpal bone two inches in diameter. We took them home on the handcar" [YPM acc. no. 1237].
71. These were of a camarasaurus (YPM 5869]).
72. *Camarasaurus grandis,* acc. no. 1243, box 6.
73. The northern shrike *(Lanius excubitor)*.
74. YPM 5869.
75. YPM 4861.
76. Same individual as scapula YPM 5869, but this specimen broke up on collection and was not sent to Yale. Quarry no. 3 was near alkali springs that rotted the bones.
77. YPM acc. no. 1245.
78. Lakes's father was Rev. John Lakes.
79. YPM acc. no. 1245, box 7.
80. Probably one of Charles Darwin's reports such as *Geological Observations on the Volcanic Islands Visited During the Voyage of the HMS Beagle or Geological*

*Observations on South America*. The reports related to the voyage of the *H.M.S. Beagle* span a number of volumes on mammals, birds, and reptiles.

81. Here ten paragraphs from an English letter of June 23, 1879, replace two paragraphs of the original journal text, which reads: "Went with Ashley to explore the strata lying west of the Robber's Roost. The strata curve round beyond the roost. We followed the round rims of this curve to beyond where the Dakotah winds round; also from the railroad track and curls up into the main ridge. Ashley saw a black tailed deer in Robber's Roost under the same tree were Marsh and our party lunched. I joined Ashley on the top of the Dakotah ridge and we made a section of the bluff near the eagle's nest. The gay colors of the bluff and the peculiar erosion into the leaches was very striking. We found some hollow bones in the sauranodon horizon possibly pterodactyl. After dinner we climbed the bluff south of the station and Ashley found some crocodile bones.

   "Received pamphlet on dinosaurs from Marsh."

   The hollow bones that Lakes conjectures might be from a pteradactyl are unidentified and may no longer exist in the Yale collections. The pamphlet he received from Marsh was probably "Principal Characters of American Jurassic Dinosaurs, Part II," which appeared in the *American Journal of Science*, 3d ser., no. 17 (January 1879): 86–92.

82. Lakes painted a large panorama of Como Lake with the quarries and other landmarks (YPM print no. 2081); see Ostrom and McIntosh, *Marsh's Dinosaurs*, frontispiece.

83. Beginning in 1867, Clarence King led a party of surveyors to map and report on that part of the United States near the 40th parallel of latitude and west of the 105th meridian. King's parties were in Wyoming during 1871 and 1872 using the Union Pacific Line as a point of reference from which to map a one-hundred-mile-wide strip. See Bartlett, *Great Surveys*, pp. 141–55, 180–86.

84. Here five paragraphs from an English letter of June 24, 1879, replace six paragraphs from the original jounal text. The original reads:

   "As Mrs. Fowler, the section house cook was sick, Ashley stayed at home to cook for her.

   "Took George [Leach] with me to the snowshed east of Como and thence to No. 4 on Rock Creek. On passing over the bluff I found the *Sauranodon* bed dipping 20° and to the south.

   "The valley of Rock Creek is picturesque. The little river winds like a serpent dividing the Cretaceous bed from the red beds. The valley is about 200 feet deep and one quarter mile wide. The red beds rise like a plain of crimson behind it towards the north. The valley at this point is occupied by sauranodon beds in which we found belemnites.

   "George who was a little ahead of me startled a couple of elk out of a narrow ravine close to him. They were bucks with fine branching antlers. They leapt the ravine in fine style scouring away across the river and up over the red rocks where we lost sight of them.

"On the eaves of one of the red rocks forming the bank of the stream clouds of cliff swallows had built their clay nests and filled the air with their wings and cries.

There was a good section here showing the junction of the red Trias with the overlying Jurassic *Sauranodon* beds. Continuing to the south we came upon the Dakotah group forming as usual a high ridge to the west and south. The variegated and light clay beds standing out in strong relief in smooth rounded masses channelled and scarred by the rain erosion. On one of these rounded beds lies Quarry No. 4 at which a great deal of work was done the year before I arrived. It is in the usual *Atlantosaurus* horizon the same as No. 3. Passing on we reached the extreme eastern limit of the Como fold where the strata curl round much as they do at Robbers Roost but at a lower angle and dipping due east.

"We dined near the track under the snow fence which is located on *Sauranodon* beds."

85. Possibly William H. Reed's son, Oscar, from his marriage to Florence Bovee, who died in childbirth when Oscar was born in 1871. George Patterson, "Brief Sketch or Biography of the Life of William Harlow Reed—Pioneer Paleontologist," William H. Reed Collection no. 957, American Heritage Center, University of Wyoming.
86. William E. Carlin, William H. Reed, and Samuel W. Williston worked in quarry no. 4 during 1877, and Reed and several temporary hands worked there during 1878. The quarry produced specimens of allosaurus, apatosaurus, and camarasaurus. See Ostrom and McIntosh, *Marsh's Dinosaurs,* pp. 10–20.
87. This is a natural formation that was used as a game blind by Native Americans. For a description of these blinds, see Etienne B. Renaud, *Archaeological Survey of Eastern Wyoming Summer 1931* (Denver: University of Denver[?], 1932), pp. 24, 39.
88. Lakes had taught at Jarvis Hall in Golden since its beginning in 1869. There is mention in the 1880 Wyoming census for Laramie City of a Maria Wanless, age 44, who was born in Canada and keeping house with three sons: Frank, 16; John C., 10; and Mark W., 6. Frank and Mark were born in Colorado and John in Wyoming. There is no mention in the Wyoming census of the father.
89. At the time, Major Thornburgh reported that there was no real threat from the Utes, who were viewed as peaceful and not a problem; see his correspondence to General Crook, "Report of Indian Depredations," July 27, 1879, *Correspondence,* Military Records Division, National Archives, reproduced in Val J. McClellan, *This Is Our Land, Vol. 1* (New York: Vantage, 1977),

pp. 217–24. For a contemporary newspaper account that discounts the Indian scare, see "North Park Mines: Return of Dr. William Davis and J. Kaufman—What They Saw and Heard," *Cheyenne Daily Sun*, June 27, 1879, p. 4.

90. Big sagebrush *(Artemisia tridentata)* can reach a height of ten feet but it begins as a seedling about six inches tall.
91. Possibly YPM drawing no. 275-C.
92. The vertebra is YPM acc. no. 1252. The tooth is YPM acc. no. 1238.
93. The sauropod was *Diplodocus longus*. Its teeth were published by Marsh in *The Dinosaurs of North America*, United States Geological Survey, 16th Annual Report 1894–95 (Washington, D.C.: GPO, 1896), p. 166, pl. 26, fig. 2; and also in *American Journal of Science*, 3d ser., no. 27 (January 1884): 168, pl. 4, fig. 2.
94. This may be the watercolor mentioned in note 81.
95. The site may be mammal quarry no. 9.
96. Again, likely William H. Reed's son, Oscar.
97. Probably the brown-headed cowbird *(Molothus ater)*.
98. To be on alert and watchful. "Qui vive" literally means "live who," a sentinel's challenge to determine whose allegiance a person has, that is, what ruler do they wish a long life.
99. The coppers are members of the subfamily *Lycaeninae*. The meadow brown was possibly *Asterocampa celtis,* which has wings decorated to resemble eyes. In Greek mythology, Argus was a hundred-eyed creature.
100. Charles Dickens' *Pickwick Papers* is a cheerful tale with many amusing anecdotes. Reed was fond of Dickens.
101. The daughter Mary Chase gave birth to was named Eva Fredericka. 1880 Wyoming Census.
102. This material is unidentified among the Peabody collections.
103. This is the first explicit reference to quarry no. 9, which produced numerous Jurassic mammal specimens. Until the 1970s these specimens were the best collection for understanding Mesozoic mammals.
104. The teeth were from either dryosaurus or camptosaurus.
105. Marsh called the locations sauranodon quarries 3 and 4.
106. YPM acc. no. 1256.
107. Probably Wilson's phararope *(Phararope tricolor)* and the American avocet *(Recurvirostra americana)*.
108. In the original journal, the entry for Monday, July 14, precedes that of Sunday, July 13.
109. The phragmocone is the conical-chambered internal skeleton of a belemnite fossil.
110. YPM acc. no. 1256 was shipped July 15 from Como. Box 9 contained these specimens.

111. Lakes indicated that this entry was for July 15, the same as the previous entry. His dates are off by a day for the rest of the month of July. The corrections will be made silently here.
112. *Allosaurus fragilis* (YPM 1867).
113. Probably YPM acc. no. 1249, a can with jaws in it.
114. Possibly YPM watercolor #275-P, Geological Sections Near Como Bluff.
115. This was the discovery of quarry no. 10, which held the holotype specimen for *Brontosaurus excelsus* (YPM 1980), now known as *Apatosaurus excelsus*. The skeleton discovered in this quarry is now on display at the Yale Peabody Museum.
116. Probably the greater yellowlegs *(Tringa melanoleuca)* or the lesser yellowlegs *(Tringa flavipes)*.
117. The visit of Edward Drinker Cope to Como was a cause of considerable activity upon the part of the collectors working for Marsh. They worked to pack up fossils and secure their sites. Material from quarry no. 1 ½was in YPM acc. no. 1268. See Ostrom and McIntosh, *Marsh's Dinosaurs,* pp. 26–30. Lakes's account in his journal is more favorable toward Cope than the letter he sent to Marsh, which describes him as the *"Monstrum horrendum"* (*Marsh Papers,* August 11, 1879, reel 10, frame 611).
118. Cope was correct about the Florissant Fossil Beds.
119. Dr. John Ryder prepared life-size drawings of a skeleton of a *Camarasaurus supremus* for Cope sometime after the spring of 1878. They were approximately sixty-seven feet long and are now preserved at the American Museum of Natural History in New York. Cope had access to part of the vacated 1876 Centennial grounds at Fairmont Park in Philadelphia.
120. There are no Jurassic mammals with that style of teeth. It is not known what these teeth were.
121. This was quarry no. 12. The fossil specimens were of *Stegosaurus ungulatus* (YPM 1853). The femur bone was eventually sent to Yale in a variety of pieces, some retrieved from the dump.
122. The bones of *Stegosaurus ungulatus* (YPM 1853) at quarry no. 12 were completely disarticulated and scattered. They comprised the rear of the skull, vertebrae from all parts of the column, a humerus, a number of dermal plates, a hind leg, two ischia and, most importantly, eight tail spines. Most species of stegosaurus had only four. Did this species have eight spines or was there the tail of a second individual in the Quarry? The debate continues to this day. There is one bone illustrated in the *Marsh Papers,* August 11, 1879, reel 10, frame 612.
123. YPM acc. no. 1268, box 11.
124. The northern harrier *(Circus cyaneus hudsonius),* also known as the American harrier or marsh hawk.

125. Actually it was the left metatarsal of allosaurus (YPM 4679, acc. no. 1260).
126. All these bones were from YPM 1981, the holotype specimen of the creature Marsh named *Brontosaurus amplus* = *Apatosaurus excelsus,* the holotype specimen YPM 1981. The phalanx bone was metacarpal two.
127. In Exodus 7:25–29, an infestation of frogs was the second of ten plagues that afflicted the Egyptians who had enslaved the Israelites.
128. These were *Baptanodon* bones (YPM acc. no. 1264), shipped August 25, 1879.
129. From *Apatosaurus* sp. (YPM 4832).
130. Acc. nos. 1282 and 1284 were sent toward the end of the month.
131. These bones were from the herbivorous iguanodontid, *Camptosaurus dispar,* described originally as *Camptonotus dispar*. Camptosaurus had claws similar to Iguanodon.
132. In late August both Reed and Lakes offered their resignations to Marsh apparently after an argument. In Reed's letter to Williston, August 20, 1879, he states that the major cause was Lakes's unwillingness to shoulder, in Reed's opinion, his share of the work and Lakes's attitude about being better educated. In Lakes's letter, August 22, 1879, he offered to resign and stated that Reed was the essential person for conducting Marsh's operations at Como. Lakes mentions that he spent time on his drawings at Marsh's request. Marsh apparently telegraphed and urged Reed to stay and Lakes offered to work by himself or do sketches. *Marsh Papers,* reel 10, frames 629–630, 633; reel 13, frames 846–48.
133. Probably the short-horned lizard *(Phrynosoma douglassi)*.
134. The election was for representatives to the Wyoming Territorial Legislature. Carbon County elected two members to the Territorial Council and four members to the Wyoming House of Representatives. The Republican slate was returned from the county. "Governor's Proclamation," *Cheyenne Daily Leader,* August 28, 1879; "Election Results," *Cheyenne Daily Leader,* September 4, 1879.
135. Likely the ferruginous (gray) hawk *(Buteo regalis)*.
136. In 1877 Marsh first named the gigantic sauropod (YPM 1835) that Lakes and Beckwith discovered at Morrison: *Titanosaurus montanus*. As explained earlier in "History and Synonymy of Some Jurassic Dinosaur Species," the generic name being preoccupied, Marsh changed it to *Atlantosaurus* as *A. montanus*. This species is indeterminate. See "Notice of a New Gigantic Dinosaur," *American Journal of Science,* 3d ser., no. 14 (1877): 87–88.
137. This is quarry 1A, about a mile from Robber's Roost, not quarry 1½, which Reed worked mainly, located north of Lake Como.
138. Probably a work crew who laid railroad track.
139. The fight between Major Thomas Tipton Thornburgh of Fourth U.S. Army Infantry and Chief Colorow and the Utes occurred on September 29, 1879.

That same day, Chief Douglas and another band of Utes attacked the Milk River Indian Agency killing the Indian agent, Nathan Cook Meeker, and capturing his family. See pp. 138–39.

140. The summer and fall of 1879 were a period of drought, and the numerous forest fires, started by accident or design, were all blamed on the Utes. There was a popular demand by whites in Colorado that "The Utes Must Go." For a contemporary account, see "Still At It," *Cheyenne Daily Sun,* September 23, 1879, p. 1.

141. Lakes apparently had submitted sketches to this English illustrated news publication. Although no illustrations from the 1879 Graphic bear Lakes's name, the April 19, 1879, issue has two sketches of "Cattle in the Far West" based on sketches by Albert H. Leith (pp. 389–90). The June 28 issue has a series of scenes "Sketches in the Territories of the United States" by Charles S. Peach (pp. 619, 620, 638,). Lakes had earlier published a sketch of the Morrison, Colorado, dig (*Graphic,* April 20, 1878, p. 392).

142. Quarry 1A produced the holotype specimen of *Camptosaurus amplus* (YPM 1879).

143. Possibly including the YPM watercolor #275-M, Geological Section, Robbers Roost.

144. These were the hind foot of a *Camptosaurus amplus* (YPM 1879).

145. These were camarasaurus bones (YPM 4633). The use of plaster of Paris as a method of preserving bones had only began to be used. Williston was experimenting with paper and paste during 1877. See Shor, *Fossils and Flies,* p. 89.

146. Possibly YPM watercolor #275-K, Looking East from *Stegosaurus* Quarry, Robbers Roost.

147. YPM watercolor #2083, Winter Quarters on the Bank of Lake Como. See Ostrom and McIntosh, *Marsh's Dinosaurs,* p. 33, fig. 10.

148. Quarry 2B never produced anything of consequence. Its location is unknown.

149. Lakes discovered a sauropod camarasaurus (YPM 4633); the skull has not been studied and identified.

150. YPM 1853.

151. YPM acc. no. 1346.

152. O. C. Marsh, "Principal Characteristics of American Jurassic Dinosaurs, Part III," *American Journal of Science,* 3d ser, no. 19 (1880): 253–59.

153. In the early morning of March 12, 1880, the Union Pacific Railroad's train no. 4 from the west collided with train no. 3 from the east during a snow storm at Red Desert Station fifty miles west of Rawlings. Train no. 3 was parked waiting for no. 4 to go to a siding. Two engines of each train were wrecked. Mathew Martin, lead engineer on train no. 4, was killed. "Railroad Collision," *Cheyenne Daily Sun*, March 13, 1880, p. 4; "Fatal Collision of Trains," *Cheyenne Daily Leader,* March 13, 1880, p. 4.

154. Frederick Brown (1842–90), immigrated from Bavaria and worked as a collector for Marsh from 1880 to 1889 and was responsible for the most detailed maps and quarry diagrams from Como. See Ostrom and McIntosh, *Marsh's Dinosaurs,* pp. 34–47; 1880 Wyoming Census.
155. Miser was a station about twenty miles east of Como after Rock Creek and forty-five northwest of Laramie. Lookout was twenty-eight miles northwest and Cooper Lake twenty-four miles from Laramie. The Jelm Mountains are thirty miles southwest of Laramie.
156. The Union Pacific built the Laramie Rolling Mills in 1874 to produce rails, plates, etc. used for its railroad and mining operations. During a quarter century its capacity increased such that by 1901 it employed 350 men and consumed thirty-eight tons of coal per day when fully operational. A fire destroyed the mill in 1910. See *Laramie—Gem City of the Plains* (Dallas: Curtis Media, 1987), pp. 266–67.
157. Sherman is about twenty-six miles east of Laramie.
158. This may refer to the as yet unlocated quarry where the Lower Cretaceous armored dinosaur *Nodosaurus textilis* was found. This is an important clue to its relocation.
159. The Snail Shell Quarry has not been located.
160. The Two Row Bone Quarry has not been located.
161. Other than an illustration of a section of bluff one mile east of Como, the journal ends here. Lakes sent home letters recounting events at Como during May-June 1879, and he included copies at the back of this journal. See note 2, above.

# Bibliography


Bartlett, Richard. *Great Surveys of the American West*. Norman: University of Oklahoma, 1962.

Breck, Allen Du Pont. *The Episcopal Church in Colorado, 1860–1963*. Denver: Big Mountain Press, 1963.

Caldwell, W. G. E. *The Cretaceous System in the Western Interior of North America*. Waterloo, Ontario: University of Ontario, 1975.

Carroll, Robert. *Vertebrate Paleontology and Evolution*. New York: Freeman, 1988.

Cope, Edwin Drinker. *Report of the United States Geological and Geographical Survey of the Territories: The Vertebrata of the Cretaceous Formations of the West Volume II*. Washington, D.C.: Government Printing Office, 1875.

Daniels, Bettie Marie, and Virginia McConnell. *The Springs of Manitou*. Manitou Springs, Colo.: Manitou Springs Historical Society, 1982.

Department of the Interior. *United States Geological and Geographical Survey of the Territories*. Annual Report. Washington, D.C.: Government Printing Office, 1873, 1874, 1875, 1875.

Egan, Ferol. *Frémont: Explorer for a Restless Nation*. Garden City, N.Y.: Doubleday, 1977.

Emmitt, Robert. *The Last War Trail: The Utes and the Settlement of Colorado*. Norman: University of Oklahoma Press, 1954.

Engelbretson, Doug. *Empty Saddles, Forgotten Names*. Aberdeen, S.D.: North Plains Press, 1982.

Flynn, Norma. *Early Mining Camps of South Park*. N.p.: n.p, 1953.

Gustafson, Carl Stanley. "History of Vigilante and Mob Activity in Wyoming." Thesis, Dept. of History, University of Wyoming, 1961.

Hallam, Anthony. *Jurassic Environments*. London: Cambridge University Press, 1975.

Harland, Walter Brian, et al. *A Geologic Time Scale*. Cambridge: Cambridge University Press, 1990.

Jones, Olive M. *Bibliography of Colorado Geology and Mining From Earliest Explorations to 1912*. Colorado Geological Survey Bulletin 7. Denver: Smith, Brooks, 1914.

Lamar, Howard. *The Reader's Encyclopedia of the American West*. New York: Harper & Row, 1977.

Lanham, Url. *The Bone Hunters*. New York: Columbia University Press, 1978.

LeRoy, L. W., and R. J. Weimer. *Geology of the Interstate 70 Road Cut Jefferson County, Colorado*. Golden: Department of Geology, Colorado School of Mines, 1971.

McClellan, Val J. *This Is Our Land*. New York: Vantage Press, 1977.

Mudge, Melville R., and Dorothy L. Mudge. *The Life of Benjamin Franklin Mudge in Kansas*. Lakewood, Colo.: privately published, 1990.

Nelson, Ruth Ashton. *Handbook of Rocky Mountain Plants*. Tucson: Dale Stuart King, 1969.

Osborn, Henry Fairfield. *Cope: Master Naturalist*. Princeton, N.J.: Princeton University Press, 1931.

Ostrom, John, and John McIntosh. *Marsh's Dinosaurs*. New Haven, Conn.: Yale University Press, 1966.

*Othniel Charles Marsh Papers*. Reels 10 and 13. New Haven, Conn.: Yale University Sterling Library Special Collections.

Patterson, George. "Brief Sketch or Biography of the Life of William Harlow Reed—Pioner Paleontologist." William H. Reed Collection, no. 957, American Heritage Center, University of Wyoming.

Patterson, Richard. *Historical Atlas of the Outlaw West*. Boulder, Colo.: Johnson Books, 1985.

Patterson, Richard. *Wyoming's Outlaw Days*. Boulder, Colo.: Johnson Books, 1982.

Platte, Robert. *The Dinosaur Hunters: Othniel C. Marsh and Edward D. Cope*. New York: David McKay, 1964.

Rhodes, Frank, Herbert Zim, and Paul Shaffer. *Fossils: A Guide to Prehistoric Life*. New York: Golden Press, 1962.

James Grafton Rodgers Collection, Colorado Historical Society.

Romer, Alfred Sherwood. *Vertebrate Paleontology*. Chicago: University of Chicago, 1966.

Schuchert, Charles, and Clara Mae LeVene. *O. C. Marsh, Pioneer in Paleontology*. New Haven, Conn.: Yale University Press, 1940.

Simmons, Virginia McConnell. *Bayou Salado: The Story of South Park*. Revised edition. Boulder, Colo.: Fred Pruett Books, 1992

Spalding, David. *Dinosaur Hunters: Eccentric Amateurs and Obsessed Professionals*. Rocklin, Calif.: Prima Publishing, 1993.

Sprague, Marshall. *Massacre: The Tragedy at White River*. Boston: Little, Brown, 1957.

# Index